No Age Limit

Short Articles

No Age Limit

Short Articles

Dr. Fahmi Bishay

Senior Consultant in Agricultural Policy and Investment;

Life Guest Professor at the Chinese Academy of Agricultural Sciences; Beijing, China;

Former Head of Department of Agricultural Economics, UWA, Perth, W. Australia;

Former Chief of Policy and planning Service: FAO/UN; Rome; Italy;

Former Senior Advisor at Investment Centre: FAO/UN; Rome; Italy.

With Contribution by

Dr. Lara Bishay (MD, MPH)

Pediatric Resident at Boston Children's Hospital,

Boston, USA.

Rome, Italy

INFINITY PUBLISHING

ISBN 978-0-7414-8263-1
Library of Congress Control Number: 2013902421

Printed in the United States of America

Published February 2013

INFINITY PUBLISHING
1094 New DeHaven Street, Suite 100
West Conshohocken, PA 19428-2713
Toll-free (877) BUY BOOK
Local Phone (610) 941-9999
Fax (610) 941-9959
Info@buybooksontheweb.com
www.buybooksontheweb.com

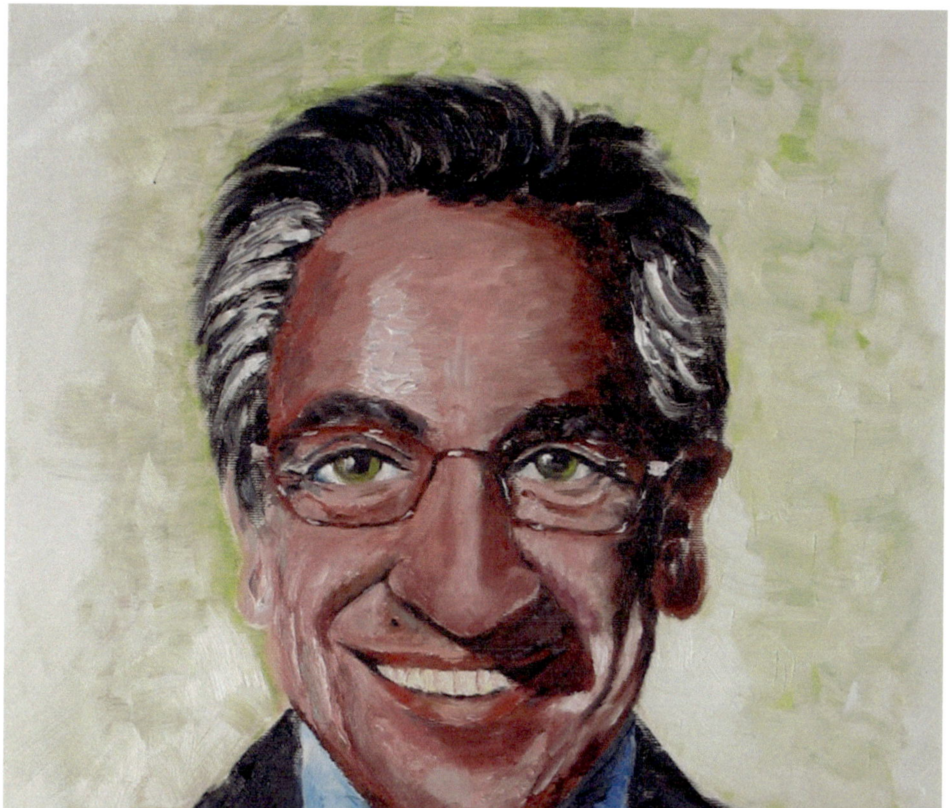

Self-portrait by the author. (Oil on canvas 70X50 cm,)

For Mona, my wife,
With love and admiration;
And
For Edoardo Francesco Bishay
Our grandson: The joy of our lives.

"Should you care to write (and only the saints know why you should) you must needs have knowledge and art and magic-the knowledge of the music of words, the art of being artless, and the magic of loving your readers."

Kahlil Gibran
Sand and Foam
(1926)

Contents

Acknowledgements

The author wishes to gratefully acknowledge the artistic skills, patience and most pleasant personality of Ms Rita Ashton for her artistic contribution and valuable suggestions in relation to the final production of the book. Rita's substantive contribution has indeed led to the production of a beautifully set and easy to follow book. For this the author is most grateful.

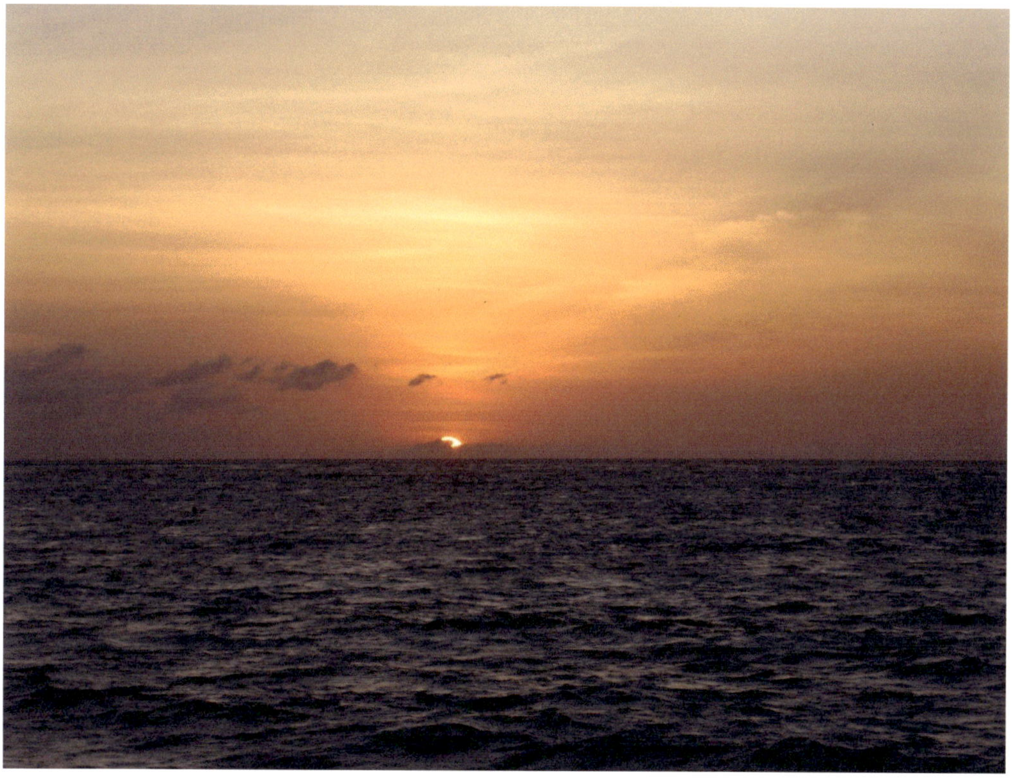

Sunrise in Bali, Indonesia: A photograph by the author; (2007).

Introduction

This book presents 20 short articles written by the author and two by his daughter: Dr. Lara Bishay. The articles are classified into six chapters covering the following subjects, respectively: (i), Art as a Hobby: Seeking Beauty; (ii), Environmental Protection; (iii), Selected World Landmarks; (iv), Sentimental Encounters, (v), The Art of Aging, and (vi), Some Concluding Remarks.

Under Chapter I, on art as a hobby: seeking beauty, there are ten articles. The first article, entitled: "No Age Limit" highlights the author's personal encounter with art as a hobby at a later stage in his life. In particular, the focus is on oil painting, Italian marble mosaics and photography. The main underpinning message is that experience has shown that art can be learned, practiced and enjoyed at any age. That is, there is no age limit to art. However, mastering art requires persistence, perseverance and hard work. And that the work of art, all sorts, must be performed with love and thanksgiving in order to be successful in reaching the hearts of the recipients. Other articles in the chapter highlight a number of lessons of experience in art that were acquired by the author while learning and practicing art as a hobby. In addition, there are two articles in the chapter highlighting the outstanding lives and works of two great artists: Paul Cézanne and Ettore R. Franz. The brief biographies of those world great artists further underscore the chapter's notion of the need of love and dedication for the success and enjoyment of art.

Under Chapter II, on environmental protection, there are three articles. The first is on the author's close encounter with the late Professor Jan Tinbergen, the first Noble Laureate in economics. The article documents the Professor's concern for the environment in the early Seventies, when the subject was not yet on the agenda of the majority of neither the world's scientists nor the economists. The second article is on the tradeoffs between economic growth and environmental protection. The third article is on application of selected lessons of experience from the second article in the agricultural sector in China; and concurrently draws further lessons of experience from the Chinese model. The discussions in the chapter highlight the intriguing paradox that: it is the process of accelerated economic growth that often leads to pollution and natural resource degradation; yet only through accelerated economic growth can environmental protection and natural resource conservation be sustainably maintained. It is further argued that it is not only accelerated economic growth that is needed, but the model of growth matters. And it is concluded that for the achievement of an everlasting longer term economic development, there is a need for an integrated approach leading to growth, equity and sustainability.

Under Chapter III, on selected world land marks, there are six articles on: the Great Pyramid of Giza; the Nile River; Coptic Icons from Egypt, Palmyra in Syria; Egyptian Obelisks in Rome and Swaminarayan Akshardham in New Delhi. The message here is that with hard work, love, dedication and determination there is no limit to what man can achieve. The Obelisks' article in this chapter was written by Dr. Lara Bishay. This article, in particular, further demonstrates how great civilizations - in this case: the ancient Romans and the Egyptian Pharaohs- sometimes meet and produce joint master pieces in art and culture.

Under Chapter IV, on sentimental encounters, there are two articles on how love does not know spatial or temporal boundaries. One of the articles is written by Dr. Lara Bishay, entitled; "A Necklace of Flowers in China"; on her sentimental encounter and love of Chinese children in the countryside of Sichuan, China. The other article analyses a great Italian proverb: "Il Lavoro Nobilita" (Labor ennobles you) on the value of labor. In this article, the author extended the message in this proverb by advocating that it is only when labor in performed with love, that it bestows honor on the worker.

Under Chapter V on the art of aging, there is one article entitled: vintage beauty. The article argues that: "When we age with grace, and we all can and should, we shall have faith and real strength."

The last chapter presents some concluding remarks. This chapter highlights the author's thesis that true love and thanksgiving should, indeed, underscore and cut through all the subjects discussed in the book. The message of the book is the need to work with love irrespective of the subject, trade, skill, location, space and/or time. Always we have to work with love. For only then shall we be blessed, and only then shall we produce sweet fruits, beauty and outstanding landmarks.

"Simplicity is the ultimate sophistication."

Leonardo Da Vince

Chapter I

On Art as a Hobby: Seeking Beauty

"(Beauty is) an image you see though you close your eyes; and a song you hear though you shut your ears."

The Prophet
Kahlil Gibran
(1923)

I.1. No Age Limit

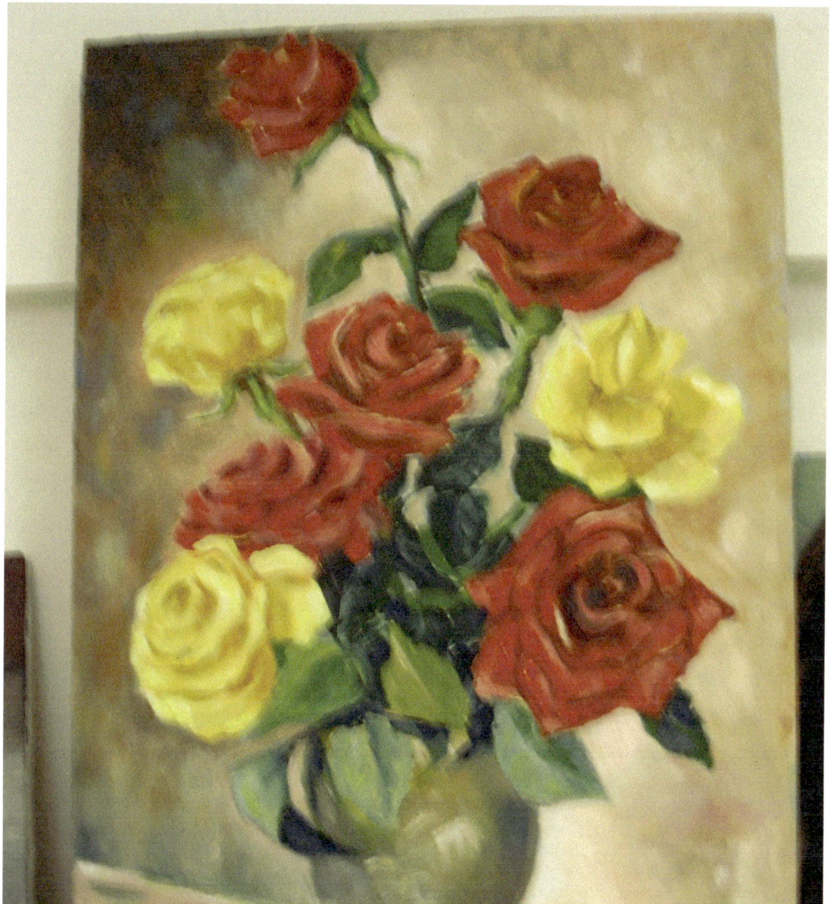

Figure 1. "A Portrait of Roses". A painting by the author. (Oil on canvas 60X40 cm.)

Introduction

During my childhood, I loved painting. However, my parents, like most, believed that a better future awaited me somewhere else. So I became an economist, a well-established economist: assumed the position of Professor of Economics at a number of Universities in Egypt, USA, Holland and Western Australia, and during the past 20 years was Chief of the Policy Analysis Service, and a Senior Adviser at the Investment Centre at FAO, Rome, Italy. ***During all those years, I never held a brush in my hand.***

However, a year prior to my retirement, my childhood dream was awakened. I subscribed in the famous ***"Artist magazine"***, participated in an art course at FAO, in Rome and started to paint with enthusiasm, determination and devotion.

My teacher was ***Mr. Bob Dickerson*** (see below), a great artist himself and a wonderful teacher. He opened my eyes on the wonderful world of oil painting. His admirable approach in teaching has been to explain the principles, but let students develop their own characters and personalities in painting. He advised us to study art, visit art galleries, appreciate the beauty in the Masters' work and practice, practice, practice. So I did. A sample of my oil paintings is herewith attached.

Upon completion of my first "reasonable" painting, I brought it home from the art class. It was a landscape of "Campo de Fiori" in Rome (oil on canvas 40 x 60 cm). My wife liked it, got it framed and hung it in a corridor between the kitchen and the bathroom. Few paintings later, she was more pleased with another piece, and hung it in a corner in the study room in place of another painting we had. With that action, I learned my wife's message: ***improve your work*** and your output could be ***"promoted"*** to compete with the good paintings already hanging on our walls in our home.

I worked very hard with commitment and determination. Now my family and friends love my work that covers all the walls in our home with the exception of two beautiful old paintings we have: a water colour painting of wild flowers by a Japanese artist who is a friend of ours and an oil painting of the ***"Kun Lun"*** mountains in China by an old Chinese artist.

So, I asked myself, "Do I have two more to conquer?" And I answered: No, I do not think so; for we love those paintings too. And we shall keep them along with my other work. As a matter of fact I felt ***honoured*** that my humble work is being hung besides those master pieces.

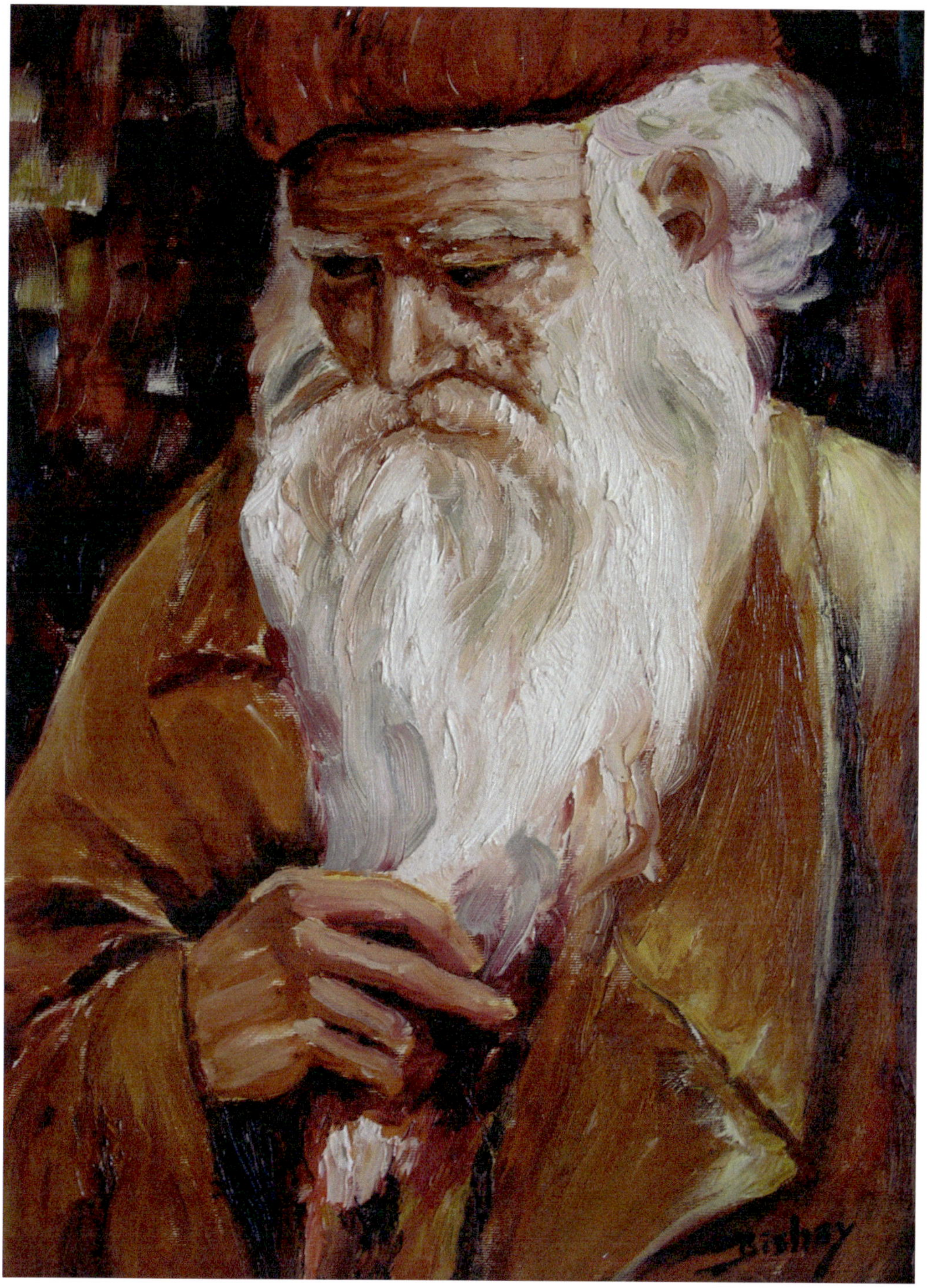

Figure 2. "Pondering life". A painting by the author.
(Oil on canvas 70X50 cm.)

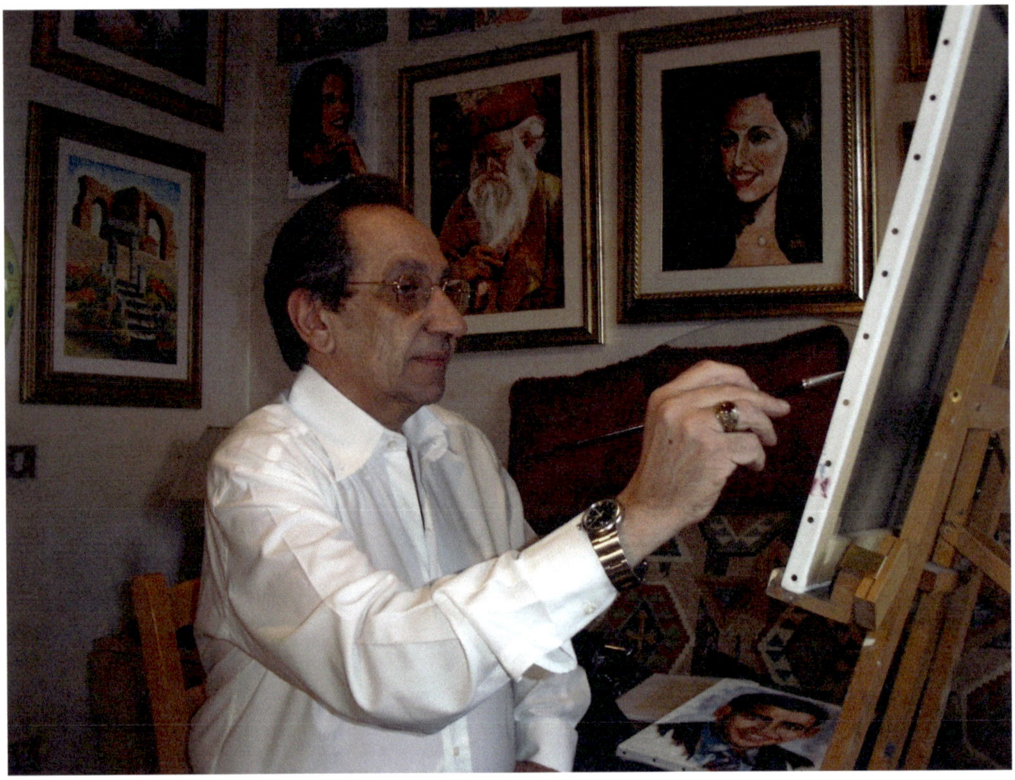

The author painting in his studio at home, Rome (2009)

On Learning Oil-painting

Bob Dickerson, my art teacher, is a very good teacher. He started the course with a number of introductory lessons on such topics as: what is oil painting about; the material including such things as the canvases, paints, colours, brushes, painting knives, oils for painting, cleaning materials, etc. Next, he focused, in a number of lessons, on how to use the material, and how to handle the oils, the paints, the brushes and the knives.

After this stage, where the students learned the basics of the painting techniques, Bob started ***the real art teaching***. Here is the interesting part. He emphasized his ***free*** style of teaching art. Free in the sense that each student was given the opportunity to choose the subject to paint and the approach of his or her painting style. That is the perception that The author liked very much about the teacher's approach. I recall that Bob Dickerson used to say, ***"I do not want to create a class of "Dickersons.*** *You (in plural) should not become copies of me. Each of you should develop his/her own specific style and character in painting".*

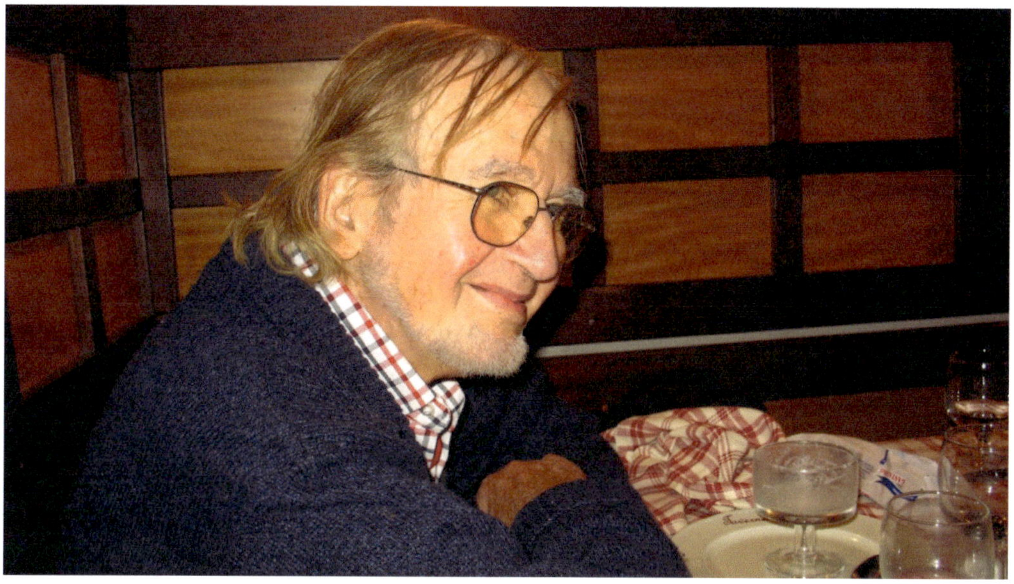

Bob Dickerson: A Great Artist; and the author's art teacher (Rome 2010).

Style and Subjects

Concurrent with the art classes, I did two important things: first, I visited as frequently as feasible the major art museums and galleries in Italy, England, the US and a number of other European countries. And I liked very much the ***"French Impressionist School"*** of paintings. In particular I have been very much fond of the works of ***Paul Cézanne, Claude Monet, Renoir and van Gogh.*** I loved the colours, the vividness of the work, the freedom of the style, and the beauty of the impressionists' paintings. At the same time, I was also very much impressed by the enormous skills of the works of the classical school. In particular, I found the works of such artists as ***Rembrandt, Caravaggio, and Michael Angelo*** to be outstanding. The treatment of ***"light"*** in Rembrandt; ***"faces"*** in Caravaggio, and the ***"human bodies"*** in Michael Angelo have indeed pushed the frontiers of "Art" into new outstanding dimensions.

My personal preferred (subjective) approach to art is that it should ***not*** copy life exactly like a "camera", for we have cameras. The impressionist school does not copy reality as such, but interpret it in a most remarkable and personal perspectives. And that is why I liked that school of art. I have, therefore, tried to belong to that school's teaching and approach.

With the above in mind, I have further acquired a "***personal art library***". I bought a relatively large number of art books (mostly on that school) with particular emphasis on the techniques of oil painting and the lives and work of the world great artists particularly the impressionists. And I have studied almost all of them. It has been a fascinating "***art journey***". The art literature enhanced my understanding and appreciation of the impressionist school of painting and the main techniques of oil painting.

As regards the subjects to paint, I found the ***"human face"*** to be one of ***God's miracles.*** No two faces are identical. The face has the character of the person, the feelings at the moment of "portraying". Even the inner feelings of the human being can be seen in the face particularly in the eyes (see the author's article on: ***stories behind the eyes,*** Section I.2. below). So, I loved the art of ***"portraits"*** and that became my most preferred subject for painting. Yet, it is perhaps for the same reasons that portraits are the ***most difficult*** and challenging form of art. For, as noted, the Artist has to go ***beyond*** the resemblance in a portrait. He/she has to dig deep in the character of the person being portrayed. I have, thus, done a good number of portraits. Some were more successful than others. It is interesting to note that according to my personal experience, the more the artists knows the person being portrayed, the higher the chance of success.

The second subject that I liked in oil painting was ***landscape*** from nature. The fascination here was due to the fact that you do ***not have to*** copy nature as such. You pick and choose. There may be twenty trees in front of you and you may pick only an "opening" couple: one on the right and another on the left: Like a curtain opening to a theatre. Then you may pick up a few more at a far distance (thus much smaller in size) to show depth, etc. In other words, you design or ***compose*** the landscape and nature becomes just a reference.

The third subject I liked was ***flowers. Flowers, particularly roses, are like "faces": No*** two roses are identical. There are "happy" roses and "sad" roses, "old" roses and "young" roses, "bright" roses and "dull" roses. Therefore, what the artist does with flowers are "***portraits of flowers***". And like in human portraits, the setting of the flowers, the colours, the foreground and the background are crucial elements for the success of the flower paintings.

How to go about an oil painting? While there is no standard answer, the author's holistic approach is explained in an article to follow (Section I.7.) entitled ***"Towards a holistic approach to oil painting"***. Further insights and examples on the art of oil painting is provided in other articles in this chapter; e.g. the article entitled: ***when a painting is finished?*** (Section I. 3).

Stages in Painting: Demonstration of an Example

The following figures show an example of key stages in oil painting demonstrated in one of the author's painting. The painting is of an Egyptian girl drinking water (not wine) from a clay pot in a hot summer night. First you sketch the subject using very light oil colour. Second you start painting, preferably from the far away elements to the closer ones. Third, you finalize the work where each brush stroke matters.

This, of course, is an over-simplification of the approach. There are ***many stages of revision*** in between, and as Bob Dickerson used to tell us, you need to ***practice, practice and practice.*** It is with practice, keen observation and never giving up that you will excel in painting and you will feel the pleasure of success and great achievements.

No Age Limit

In view of my own experience with oil painting at an advanced age highlighted above, I concluded that ***art is actually about commitment, determination and energy – not age.*** My personal experience demonstrates that there is indeed ***no age limit to art.***

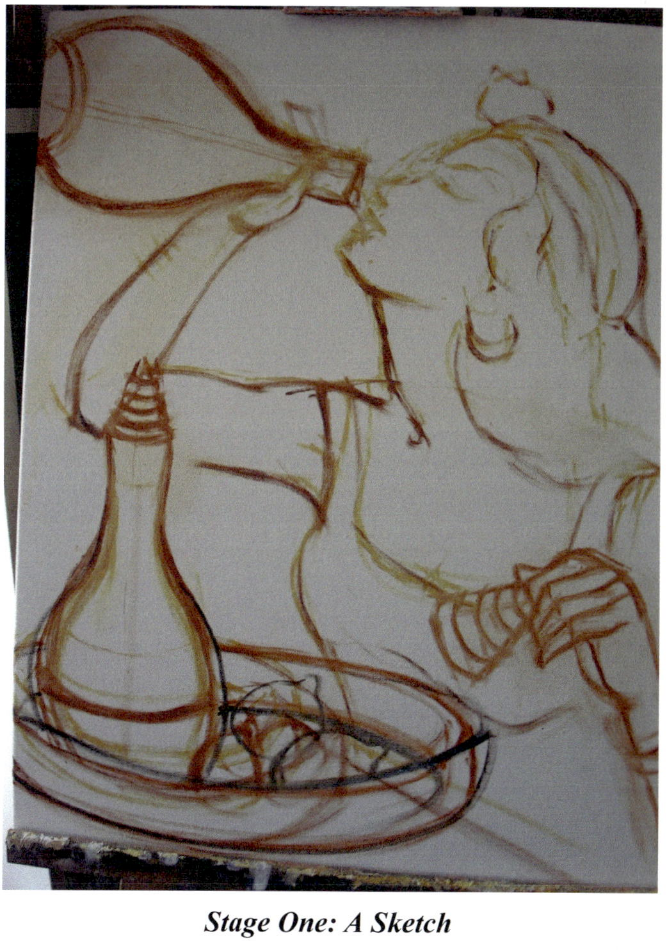

Stage One: A Sketch

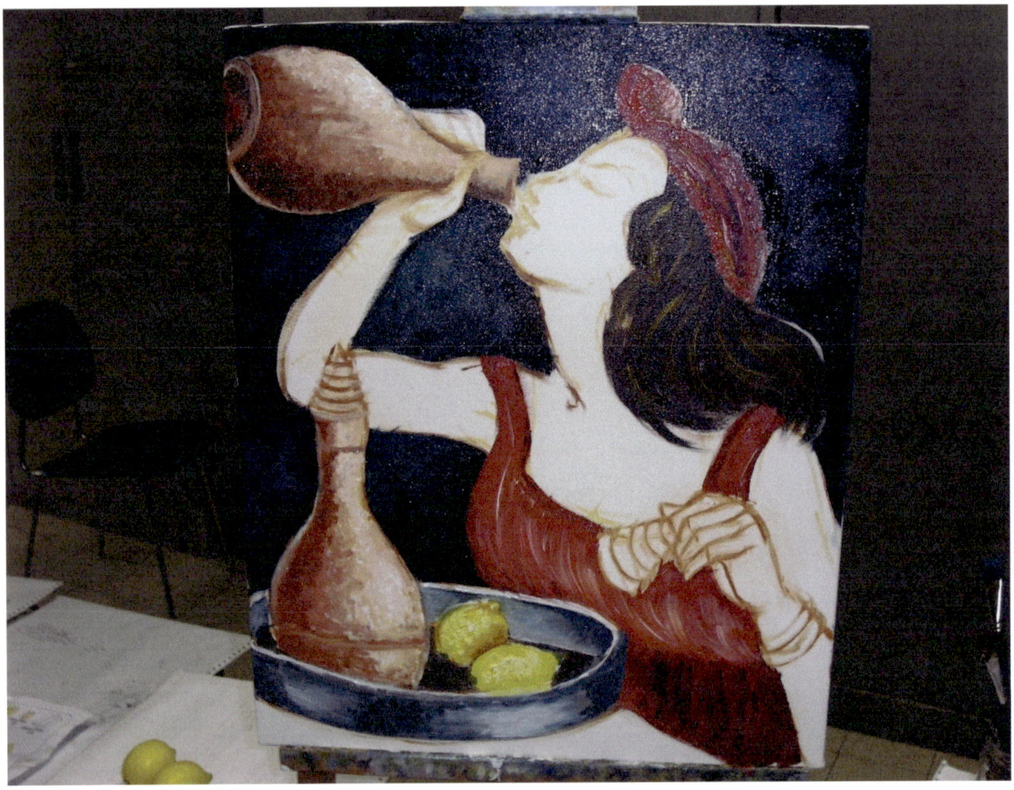

Stage Two: Half way through

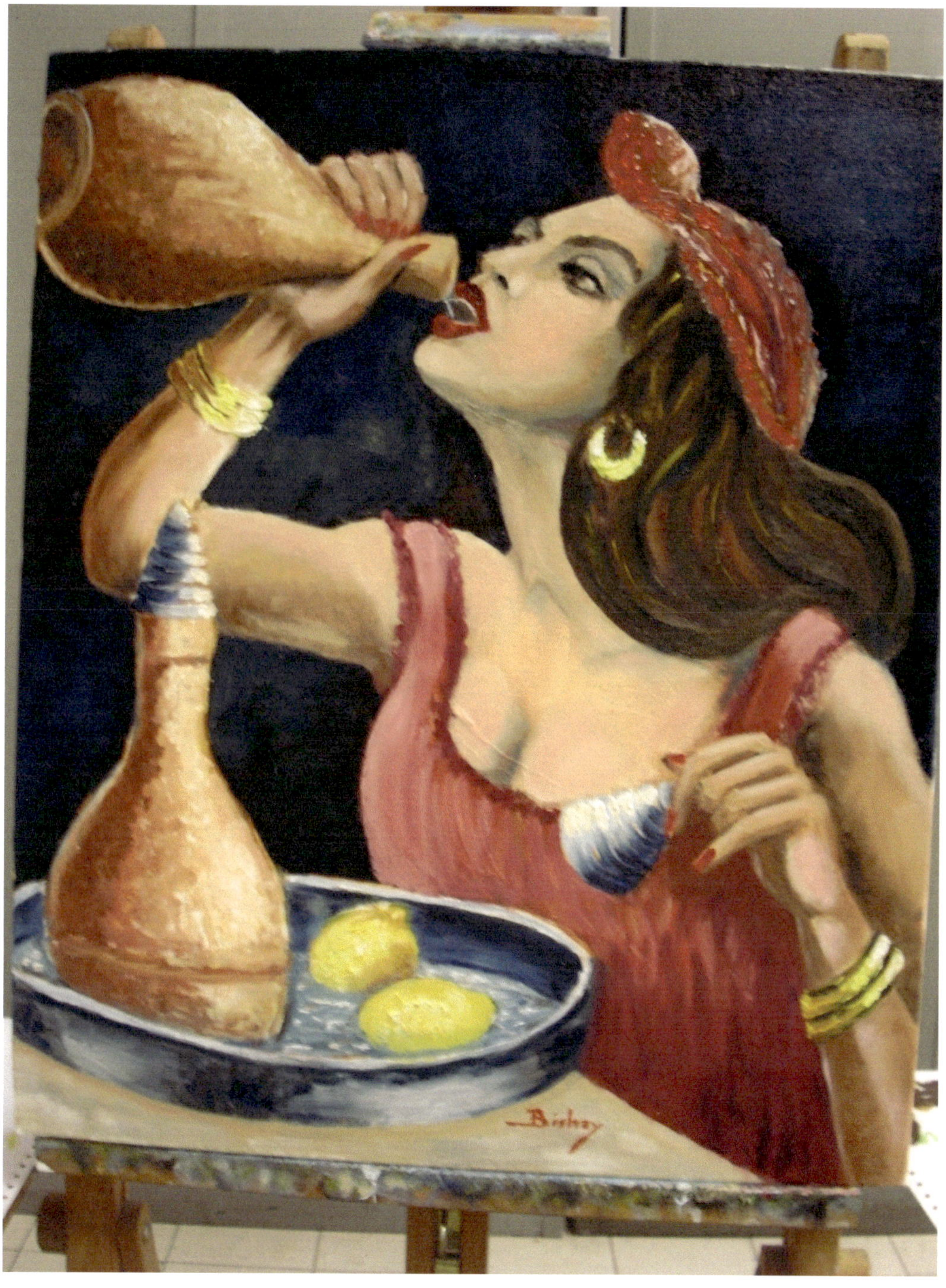

Stage Three: Final. "Thirsty": A painting by the author
(Oil on canvas: 50X40 cm)

I.2. Stories behind the Eyes

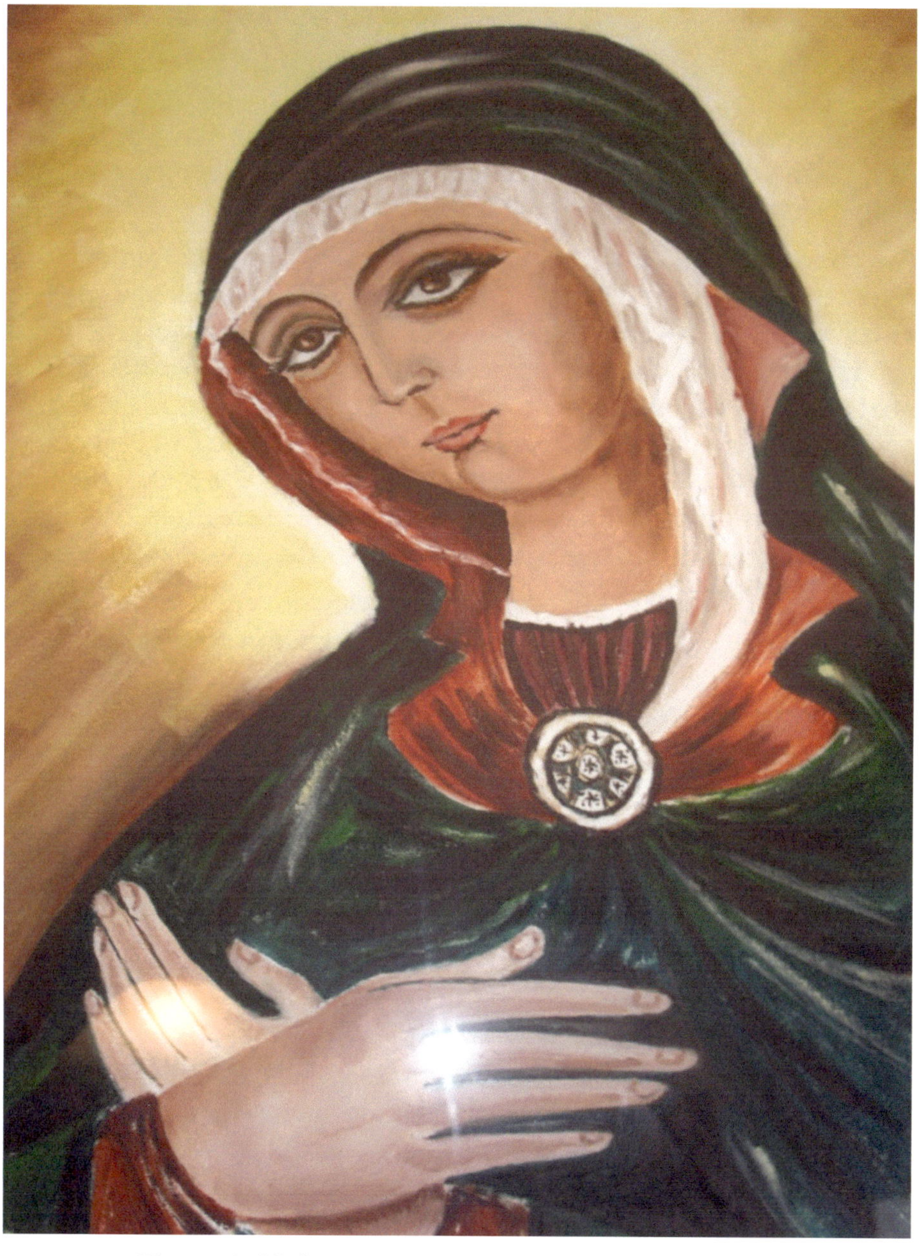

Figure 1. "Saint Mary: A Coptic Icon from the East".
A painting by the author.
(Oil on canvas 70X50 cm.)

Introduction

It is said that the ***"eyes are the windows to the soul"***. How true. When we meet people, our first encounter with them is the expression in their eyes. Eyes could be welcoming, happy, kind or smiling. Others could be angry, afraid, uncertain or puzzled. Expressions of the eyes are not only perceived in humans but also in animals. The eyes of a predator are hunting, frightening, searching, and sharp. The eyes of the prey are scared, frightened, and anxious. Eyes can thus tell stories: actual and/or imaginary.

The above perception of how eyes are perceived is very important in art. For an artist who dares to penetrate the face in a portrait, he or she has to think very hard about the eyes of his/her model before starting work. If the artist knows his/her model, then interpreting the expressions in the face to be captured on the canvas will have a higher chance of success. If the artist does not know the character of the model, he/she must try to "read" the model's eyes. The artist will then "imagine" some stories behind the eyes. And the artist will do his best to tell the model's story through the expressions of the eyes in his final product. And this applies for painting humans as well as animals.

To appreciate the above concept, the following are a few ***examples of perceived stories, true and/or imaginary, behind the eyes in selected oil paintings by the author:***

I. Praying Eyes: An old Coptic Icon from the East

The Coptic Icons played an important role in the old Egyptian Orthodox Eastern Church. The main reason was that since the Egyptian Christian converts came from the Pharaoh's culture, they had difficulty understanding Biblical teachings and their spiritual meanings, as well as the historical events that took place in the Bible and in the life of the Church. Therefore, the leaders of the Early Coptic Church introduced Icons as a means to assist the people to assimilate Christianity and its doctrine. The Characteristics of many Coptic icons follow certain symbolism that carries a spiritual message. In particular, the Icon of Saint Mary, painted by the author and its image is shown here (Figure 1.), has large and wide eyes symbolizing the ***spiritual eye*** that look beyond the material world. The Bible says ***"the light of the body is the eye: if therefore thine eye be simple, thy whole body shall be full of light" [Matthew 6:22]***. (See an article to follow on ***Coptic Icons from Egypt, Section III.3.).***

II. Eyes and Magic: The Pharaoh's "God Horus" in Old Egypt

The story behind the eyes in the painting in Figure 2 is as follows: Christianity entered Egypt in the first Centaury, actually in the year 61 A.D. by Saint Mark, The Evangelist. Saint Mark was martyred in Alexandria in the Year 68 A.D. Saint Mark wrote his Bible in Egypt during those years. His followers, the Old Egyptian Christians, known until today as the Copts, were prosecuted by the Romans. Thus most of them escaped to the Southern region of the country or Upper Egypt. A good number of them lived in some old and deserted Pharaohs' temples. Those temples were full of statues and carvings on the walls. The Copts thought that the ***"magic"*** of the Pharaohs was manifested through the ***eyes*** of those statues and paintings. They were, thus, ***afraid of those eyes***. Perhaps the reason for this belief was the ***"strong" penetrating eyes*** of those statues. The sharp eye in the attached painting of the golden head of the statue of the God Horus tries to tell this story of ***"Eyes and Magic"***.

III. Eternal Eyes: Eyes of the Sphinx

The Sphinx spoke only once. And He said: "My name is the Great Sphinx of Giza in Egypt. I am very old. I am over 4500 years old. I am highly distinguished. I have Royal Blood running in all my veins, for I have the body of a lion, (king of animals); and the Head of a Pharaoh King: (King Chephren). I am huge. I am the world's largest man-made round stone structure. My length is over 72 meters, and my height is over 20 meters. My face is four meters wide and my eyes are two meters high. My mouth is about two meters wide. Unfortunately, I lost my nose about 400 years age. It would have been more than 1.5 meters long. ***My eyes are "eternal". Against the scale of time, my eyes have been observing life from almost "minus infinity" to "plus infinity".*** My job, over the past few thousand years, has been to guard the Great Pyramids of Giza. During those years, I have witnessed many a significant event in Egypt and in the world at large. I saw wars and destruction. I saw peace and prosperity. But, I have never understood how a man can kill his brother and claims that that was a noble act conducted for the sake of peace. ***I am longing for a world with true love and eternal peace***." That is what the Sphinx said. And He became silent again. The attached painting of the Head of the Sphinx by the author (Figure 3.) tries to capture those ***"eternal eyes"***.

IV. Comforting Eyes: The Mare and the Foal

An imaginary story that lies behind this painting (Figure 4.) is as follows: The white horse is the "mother" mare painted next to her son: the brown foal. The eyes of the foal show that he was scared and frightened. The eye of the mother mare is calm and comforting for her poor son. Some of the neck hairs of the mother are placed on the foal and some on the mother's eye. As if the mother mare is saying: ***"my son foal do not worry, you are as dear to me as my eyes".*** The painting tried to capture the mother's ***"comforting eyes"***.

It will be a fascinating exercise for you, my dear reader, to search for the true stories and/or to construct mental/imaginary stories behind the eyes in various paintings, particularly in the works of the masters.

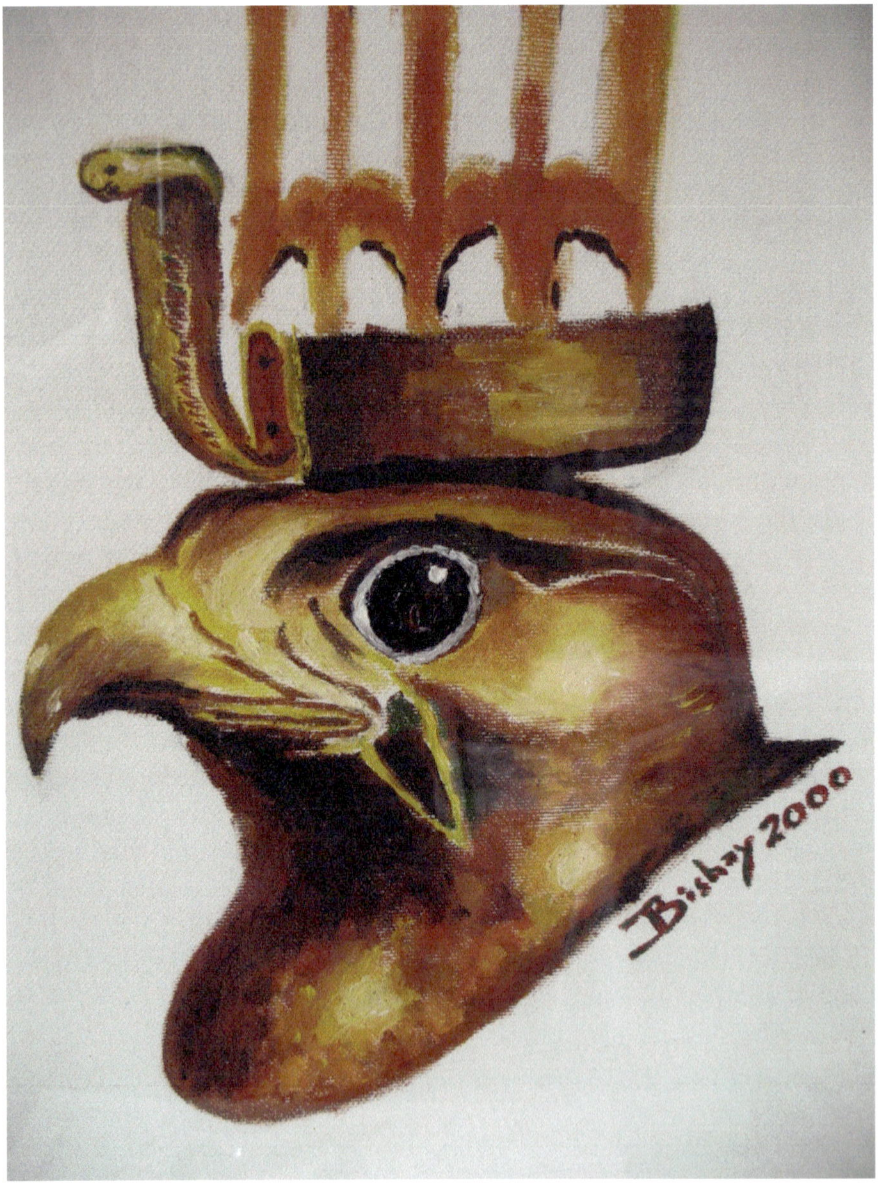

Figure 2. "Magic eyes of God Horus in Old Egypt". A painting by the author. (Oil on Canvas 60X40 cm.)

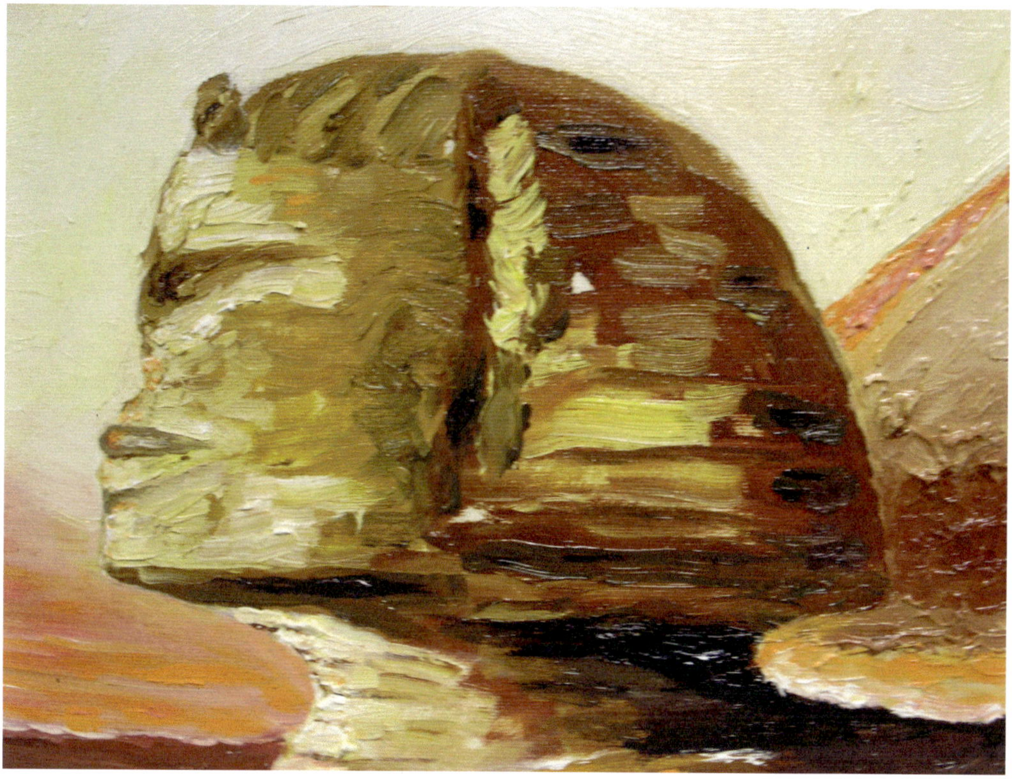

Figure 3. "Eternal eyes of the Sphinx." A painting by the author. (Oil on canvas 70X110.)

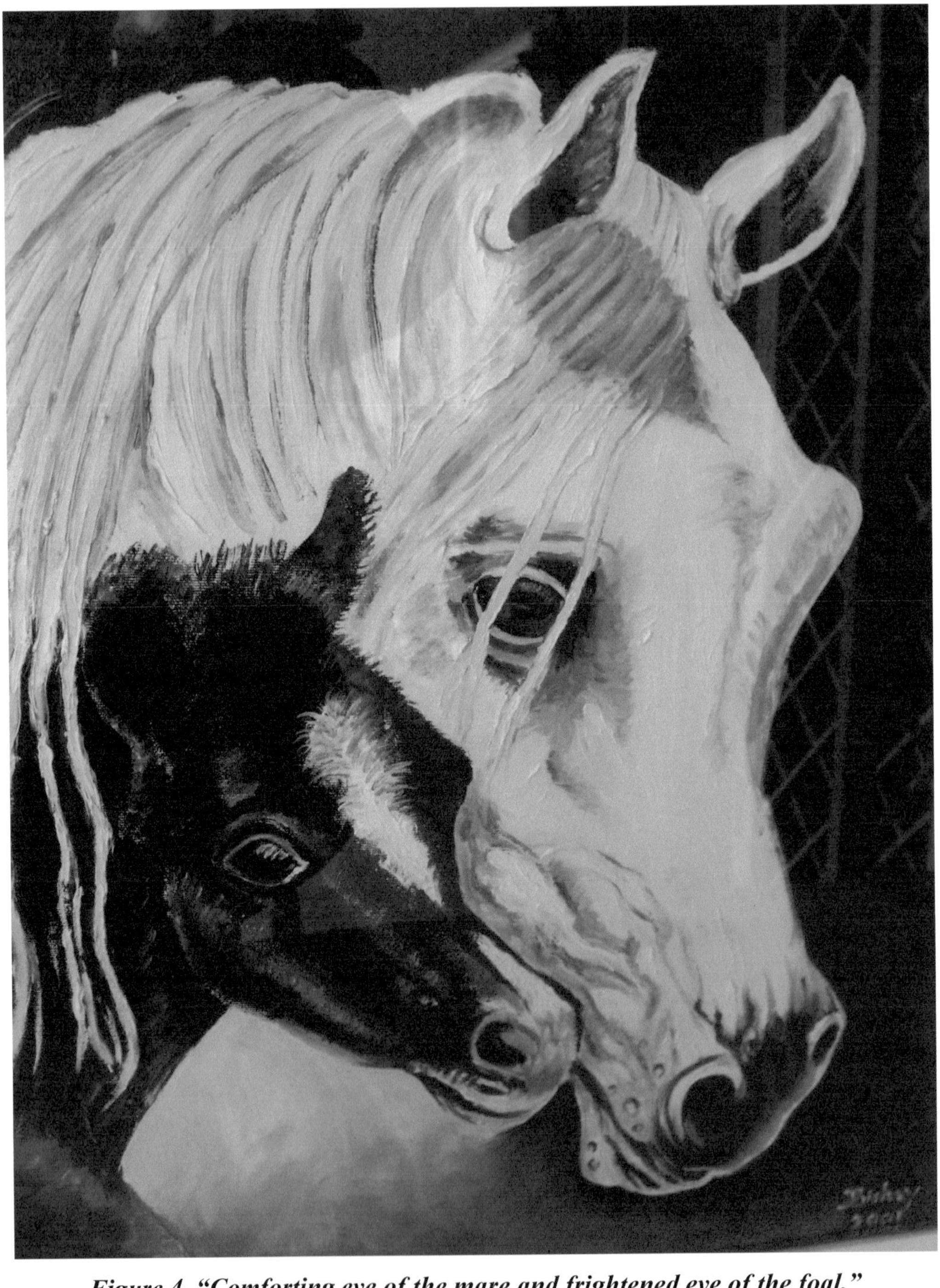

Figure 4. "Comforting eye of the mare and frightened eye of the foal." A painting by the author. (Oil on canvas 60X40 cm.)

I.3. When a Painting is finished?

Fulfillment of the Artist's Objective

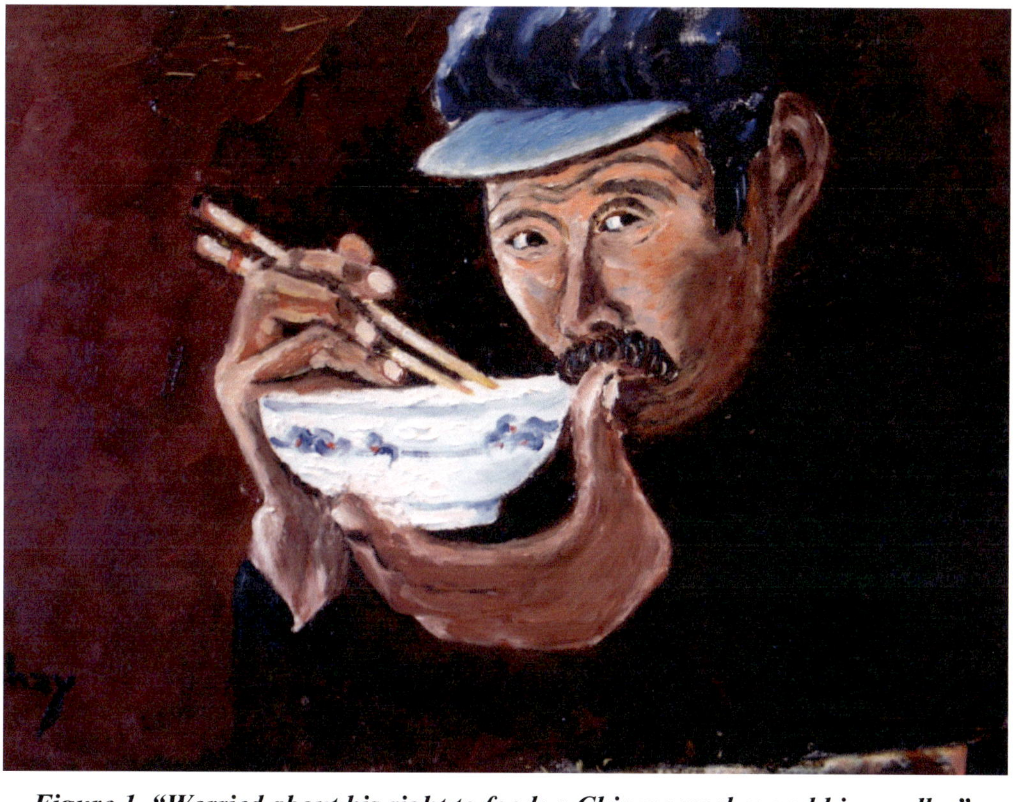

Figure 1. "Worried about his right to food: a Chinese worker and his noodles". A painting by the author. (Oil on canvas 40X60 cm).

The above question might seem obvious, and unworthy of discussion. Yet, it raises a fundamental aspect in painting, particularly for the beginners. You work on a painting, (irrespective of the media or the subject), and you keep improving, and improving and further improving your work thinking, (rather hoping), to get a better painting. However, there is an ***"optimum"*** point at which the work is at its best and that is when you have to stop for the painting is finished at that point. The problem is that you might ***not know*** when you have reached that optimum point. And not infrequently, you may surpass that point and you continue what you think is further "improvement" of the work. Actually you are damaging the work; this process in art is known as ***"over-working".*** The painting will then lose its freshness and will likely become "dull". The artist feels frustrated. He or she would then be wondering ***"why didn't I stop earlier?"*** Thus it is very important to know when you have reached the optimum point and when your painting is finished.

As an amateur painter myself, I faced this problem many a time especially at my earlier works. So I searched in the art literature trying to find the answer to the question: "when my painting is finished? And I found the answer in an outstanding art book entitled:"***Oil Painting Secrets from a Master"***[1]. The book is written by Linda Cateura about the views and suggestions, (that she labeled secrets), of the master artist: David A. Leffel.

Leffel, rightly, argues that ***before*** touching the brushes and the canvas, a painter should have a ***"concept"*** (that I call objective) to be realized from his or her painting. He says: ***"concepts involve how the painting will be read",*** he **added *"The idea (concept or objective) for a painting comes before you begin to paint. It is the artist's way of seeing things".* Once the objective is realized, you stop. The painting is finished, even if certain sections of the canvas are not completed.**

When you have an objective of the painting before you start, you will work ***towards*** that objective. For instance if you are doing a portrait for a person and you wish to express a certain psychological depth of character, you then ask yourself: how can I show this? You may decide to show the entire painting in a low key by using dim lighting, yet focusing on the strong eyes and the expressive mouth. The emphasized face will then be your "***focal point***". You may also place the lighting on one side to cast half the face in dark shadow, and so on. That is, you ***plan*** your painting beforehand. As you have a plan for your

[1] Linda Cateura, "***Oil Painting Secrets from a Master***", Watson-Guptill Publications/New York, 1995.

painting, you work towards realizing it. Thus, you would have a ***mental vision*** of the essence of your work before you start. In this example you would imagine a portrait against dark background with very strong facial features with half the face in dark shadow. So once your painting achieved that image and the strong character of your model, (your focal point), becomes clear you stop. The painting is finished.

Let us now have some ***examples*** of paintings by the author to further illustrate this approach.

Figure 1: The objective of this painting entitled ***"Worried about his right to food: a Chinese worker and his noodles"*** was to focus on the ***insecurity*** and ***worry*** of the poor Chinese worker about his right to food or his noodles. His concern is expressed in his face; that is the focal point of the painting. His worry is particularly reflected in the look in his eyes as if afraid that his treasure, the noodles, could be taken away from him. The way his "working" dirty hands are drown holding the pot of noodles and the chop sticks further enhances the objective of the painting. Once the face (including his blue working hat), the pot of noodles and his hands were in place, the painting was declared finished. One could have added some Chinese decorations in the foreground and/or in the background. But I opted, in my plan of the painting, against that idea so that not to distract from the painting's objective. The foreground and background were, thus, left plane and painted quite dark to enhance the drama.

Figure 2: The objective of this painting entitled ***"Impressionistic image of roses in a simple vase"*** was twofold: (a) to paint roses in an ***impressionistic*** approach using thick paint; and (b) to show the ***movement*** of ***light*** from left to right. The roses were thus painted in a ***rough*** fashion yet maintaining the anatomy of the flowers. The thick paint application helped providing a sort of a feeling of ***"carving"*** the roses. As regards the light movement, this objective was demonstrated on the roses, the foreground and the background. The shadow of the vase completed the light movement's concept. It should be noted that the light side of the roses faces the dark side of the background and the dark side of the roses faces the light side of the background. Once a rough image of the roses along with clear direction of the light movement was demonstrated, the painting was declared finished.

Figure 3: The objective of this painting entitled ***"Head of the Eternal Sphinx of Egypt"*** was to emphasize the ***"strong"*** character of that ancient and fabulous statue. The plan of the painting was thus to use very ***strong*** and ***contrasting colors*** of the head. In addition, the use of ***thick paint***, mostly directly applied from the tubes with virtually no mixing, enhanced achieving the objective of a strong, vivid and very effective head. The head of the Sphinx in reality does ***not*** have all those colors. It only has the yellowish color of the stone and sand. To add mystery to that eternal statue, the details of the eye were omitted, and the mouth was painted firm and expressionless. The painting was declared finished when the head of the Sphinx gave a feeling of ***gazing into the infinity***, and its rich, fresh and contrasting colors gave a feeling of ***strength, beauty and mystery.***

Figure 4: The objective of this painting entitled ***"Petunia flowers: a cool painting for a room in a hospital"*** was to make a ***cool painting*** that is relaxing to look at; and thus could be suitable for a room in a hospital. The plan was, therefore, to paint very ***light colored*** flowers; white petunias were chosen. Cool white and light blue colors were used for the foreground. And light warmer beige colors were used for the background in combination with rays of light in cool white and light blue colors. ***Thin paint*** was used to enhance the cooling and softness effect of the painting. The flowers were placed at different angels to show their anatomy. When the painting gave a ***cool, soft and relaxing*** image, the painting was declared finished.

To ***conclude***, ***before*** you start painting, you should have an objective and a plan for your painting. When your perceived objective and the essence of the plan of the painting are achieved, you stop. Your painting is finished. This approach makes the painting process ***thoughtful***, ***well guided and enjoyable.*** In addition, thinking and deciding on your objective and your plan for your painting before you start, and achieving that objective gives you a great ***feeling of satisfactory achievements.***

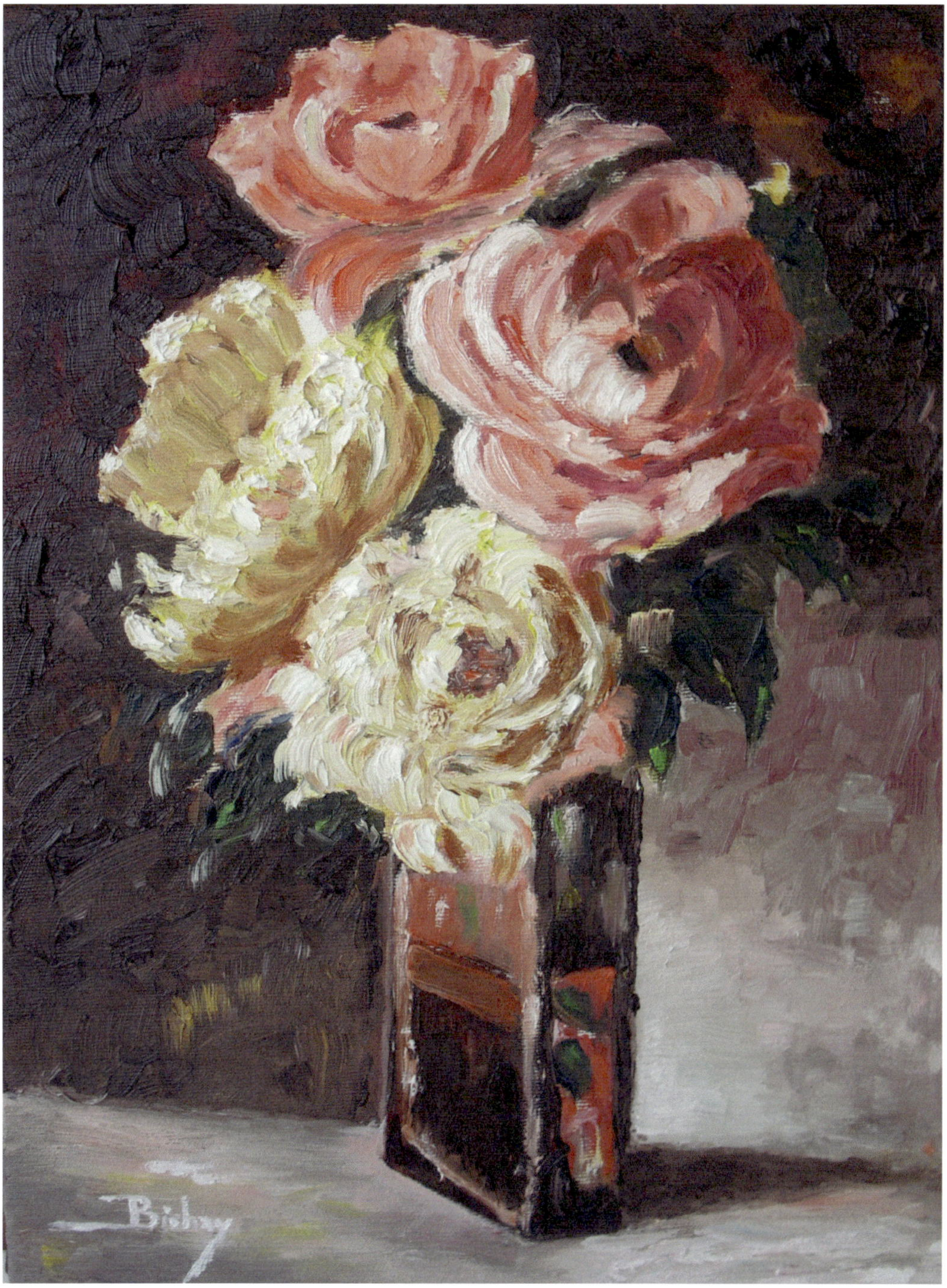

Figure 2: "Impressionistic image of roses in a simple vase".
A painting by the author. (Oil on canvas 40X30 cm).

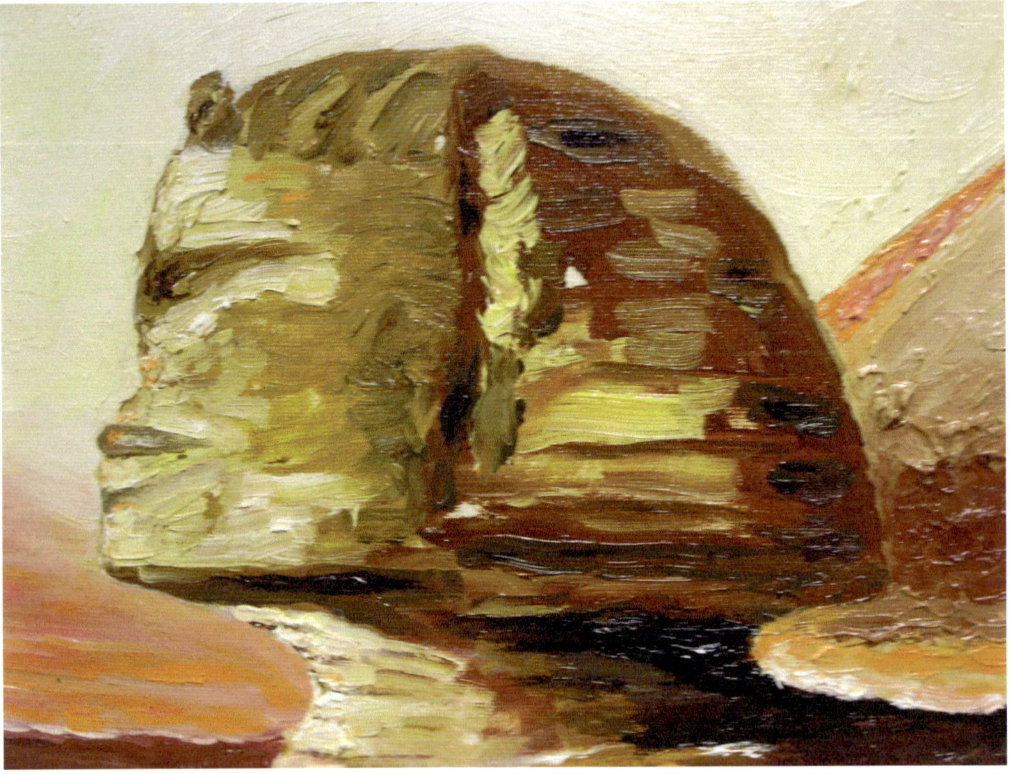

Figure 3. "Head of the eternal Sphinx of Egypt".
A painting by the author. (Oil on canvas 70X110 cm).

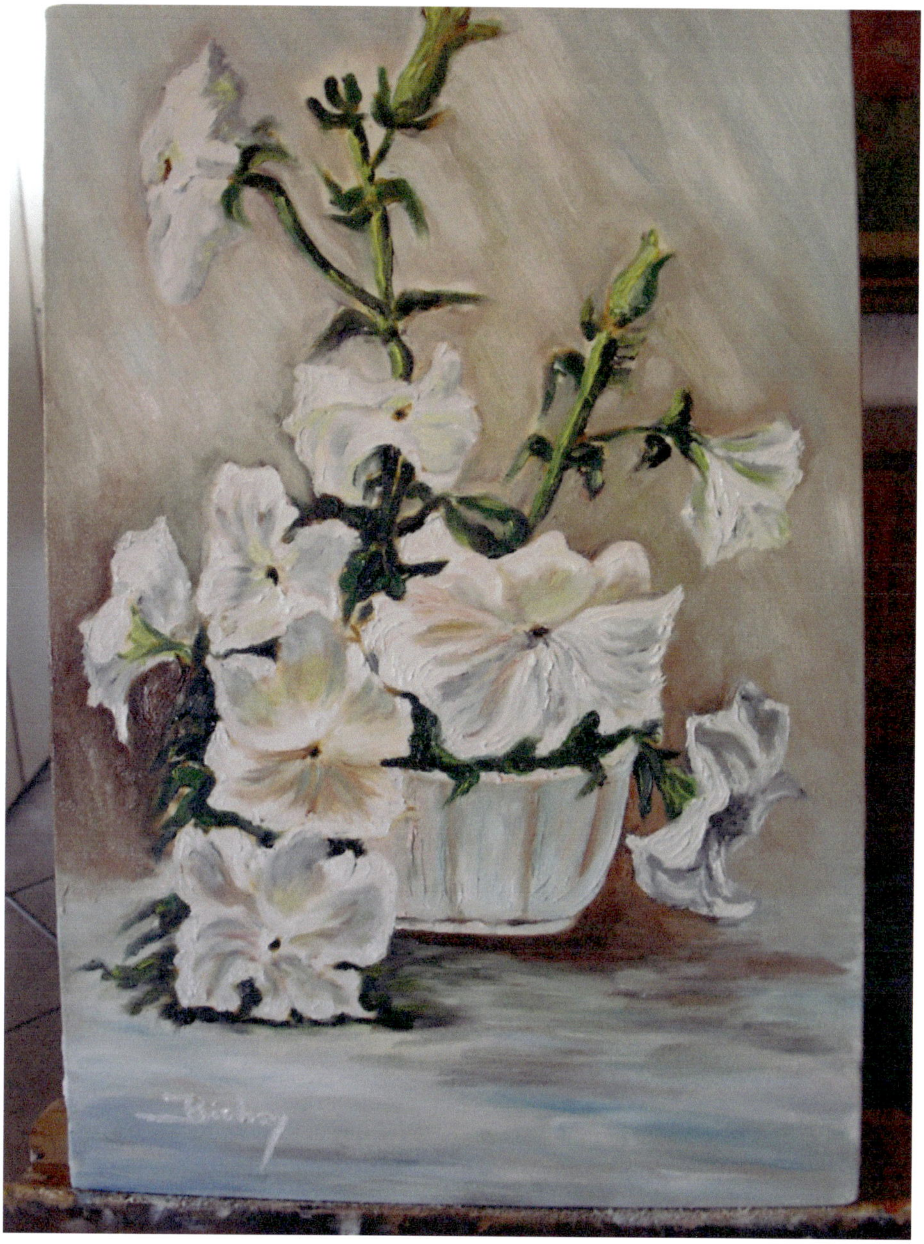

Figure 4. "Petunia flowers: a cool painting for a room in a hospital". A painting by the author. (Oil on canvas 60X40 cm).

I.4. Vladimir Madonna:

A Russian Masterpiece of Art

Vladimir Madonna, also known as the "Virgin of Tenderness", is a byzantine work of art believed to have been painted in the first century by Saint Luke the Evangelist. Vladimir Madonna is an important subject for both the Western Catholic Church and the Byzantine Orthodox Church of the East. The image depicts the Virgin Mary holding a Child: Jesus Christ. Her large eyes gaze out from the picture in sadness as she may be aware of her child's fate. Her mournful eyes cry out to us in a powerful way. The Vladimir Madonna's hands lovingly hold the Child to her. She conveys the loving emotion of the Child she holds in her arms.

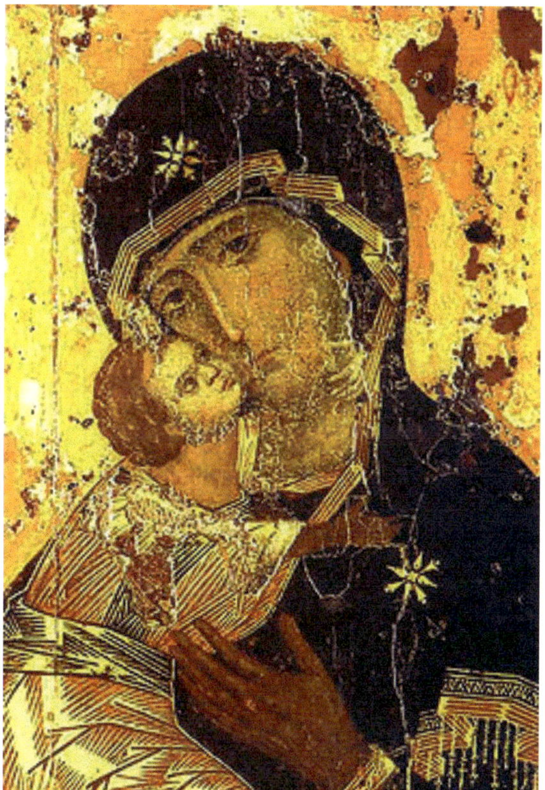

Figure 1. The Original Vladimir Madonna.

Introduction

The ***"Vladimir Madonna"*** is one of the most venerated Orthodox icons and a typical example of ***Byzantine iconography*** (Figure 1). The ***icon*** is regarded as the holy protector of Russia. The icon is currently displayed in the Tretyakov Gallery, in Moscow. Its feast day is June 3. It is believed by the church that the icon was painted by ***Saint Luke*** in the first century. The venerated image was used in celebration of coronations of tsars, elections of patriarchs, and other important ceremonies of the State. The Vladimir icon is sometimes described as expressing ***universal feelings*** of ***motherly love and anxiety for her child.***

Saint Luke: the Artist

It is believed that Saint Luke was the first icon painter. He is said to have painted pictures of the ***Virgin Mary: Vladimir Madonna*** and of Saints Peter and Paul. The tradition that Saint Luke painted icons of Mary and Jesus has been common, particularly in the Eastern Orthodox Church. The tradition also has support from the Saint Thomas Christians of India who claim to still have one of the Virgin Mary's icons that Saint Luke painted and which Saint Thomas brought to India.

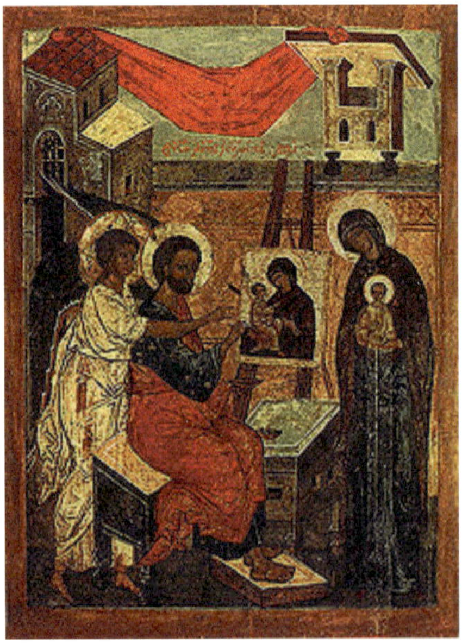

Figure 2. Saint Luke the Evangelist painting the first icon of the Virgin Mary: The Vladimir Madonna; (Accademia di San Luca in Rome).

Transportation of the Icon

Initially the icon was kept in Greece. And in about 1131 the Greek Patriarch Luke Chrysoberges of Constantinople sent the icon as a gift to the Grand Duke of Kiev. The image was kept in the Mezhyhirskyi Monastery until the Duke's son brought it to his favorite city, ***Vladimir,*** in 1155. ***"Tradition tells that the horses transporting the icon stopped near Vladimir and refused to go further. People interpreted this as a sign that the Virgin Mary wanted her icon to stay in the city of Vladimir",*** *(centre of European Russia at the time).* Thus it became known as the ***"Vladimir Madonna".*** To house the icon, the great ***Assumption Cathedral*** was built there, followed by other churches dedicated to the Virgin Mary throughout Ukraine.

In 1395, during the Tamerlane's invasion, the image was taken from Vladimir to the new capital: Moscow. The spot where people and the ruling prince met the icon is commemorated by the Sretensky Monastery. Tradition further indicates that "*Vasili I of Moscow spent a night crying over the icon, and Tamerlane's armies retreated the same day*". The Muscovites refused to return the icon to Vladimir and placed it in the Cathedral of the Domitian of the Moscow Kremlin. On June 3rd every year, the Russian Orthodox Church celebrates a holiday devoted to the Vladimir Icon of the Madonna. The icon is, indeed, one of Russia's greatest relics.

Copies of the Icon

The original icon has been repeatedly copied by many an artist for centuries. Many copies now have considerable artistic and religious significance of their own. For instance the Pskov-Pechora icon "Affection" (1524), the Zaonikiyev icon (1588), the Krasnogorsk icon (1603), the Oransk icon (1634), the Tupiches-Rostov and Florishchev icons (the 17th century), the Syrkov icon (1548) and others are celebrated in Russia on June 3 rd every year.

Inspired by the Vladimir Madonna, the author painted a Madonna and labeled it: "Bishay's Madonna" (Figure 3). As can be seen in the figure, the face of the Virgin Mary in Bishay's Madonna is not sad, but rather thoughtful and penetrating the future of her son. In addition, the expression on the face of the Baby: Jesus is almost smiling and clinging to his mother with very strong and loving eyes longing to her. In addition, bright colors were used, and the Coptic script in the background of the Bishay's Madonna was emphasized.

Concluding Remarks

Appreciating that the ***Vladimir Madonna is a masterpiece in Russian art,*** David Talbot Rice states in the latest edition of the ***Encyclopedia Britannica*** that: ***"it (Vladimir Madonna) is (an icon) of a considerable importance in the***

history of painting, for it not only is a work of outstandingly high quality but also is in a new, more human style, anticipating the late Byzantine style that flourished between 1204 and 1453."

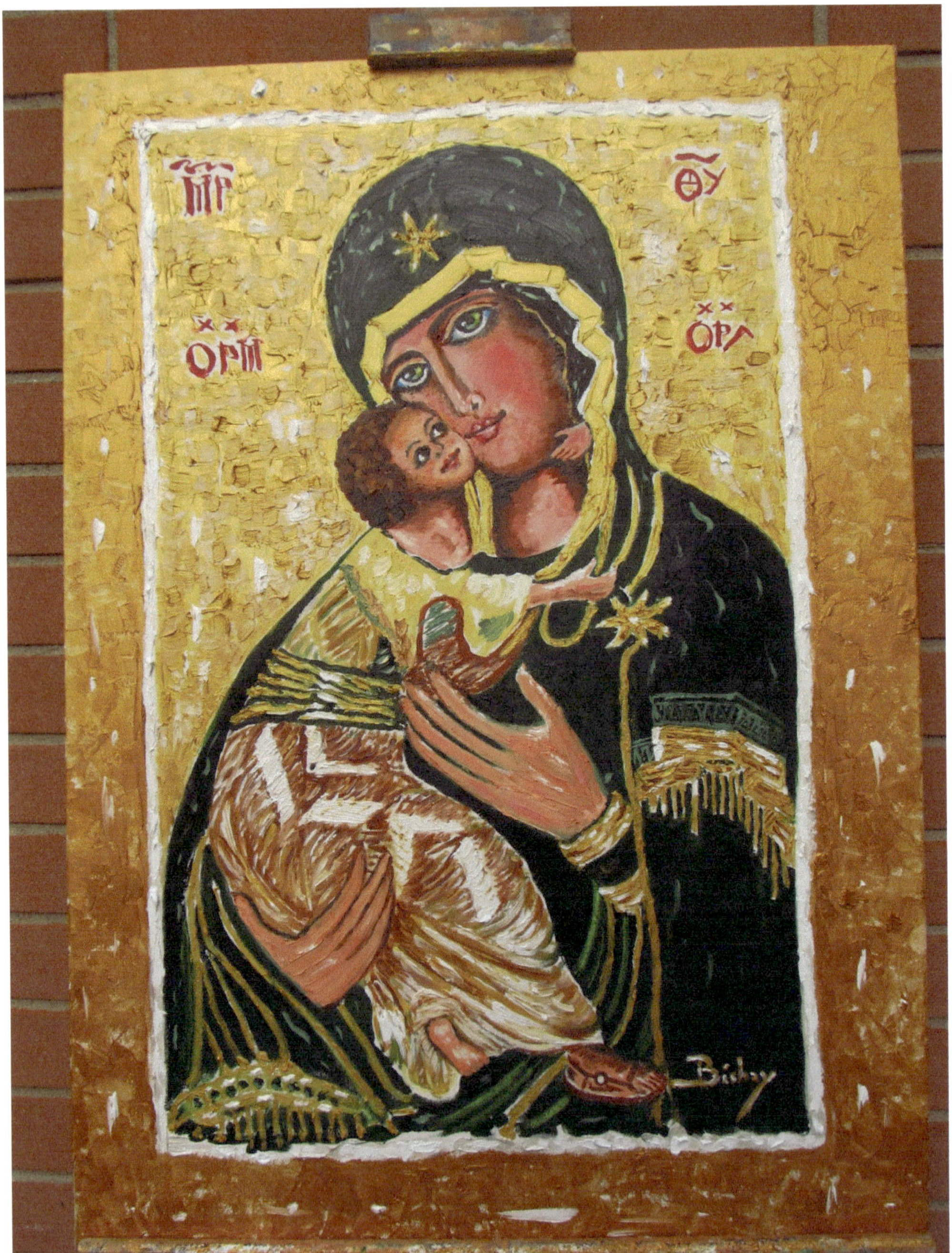

Figure 3. Bishay's Madonna: A painting by the Author inspired by Vladimir's Madonna. (Oil on canvas: 70 X 50 cm.)

I.5. Paul Cezanne:

"I would like to conquer Paris with an apple"

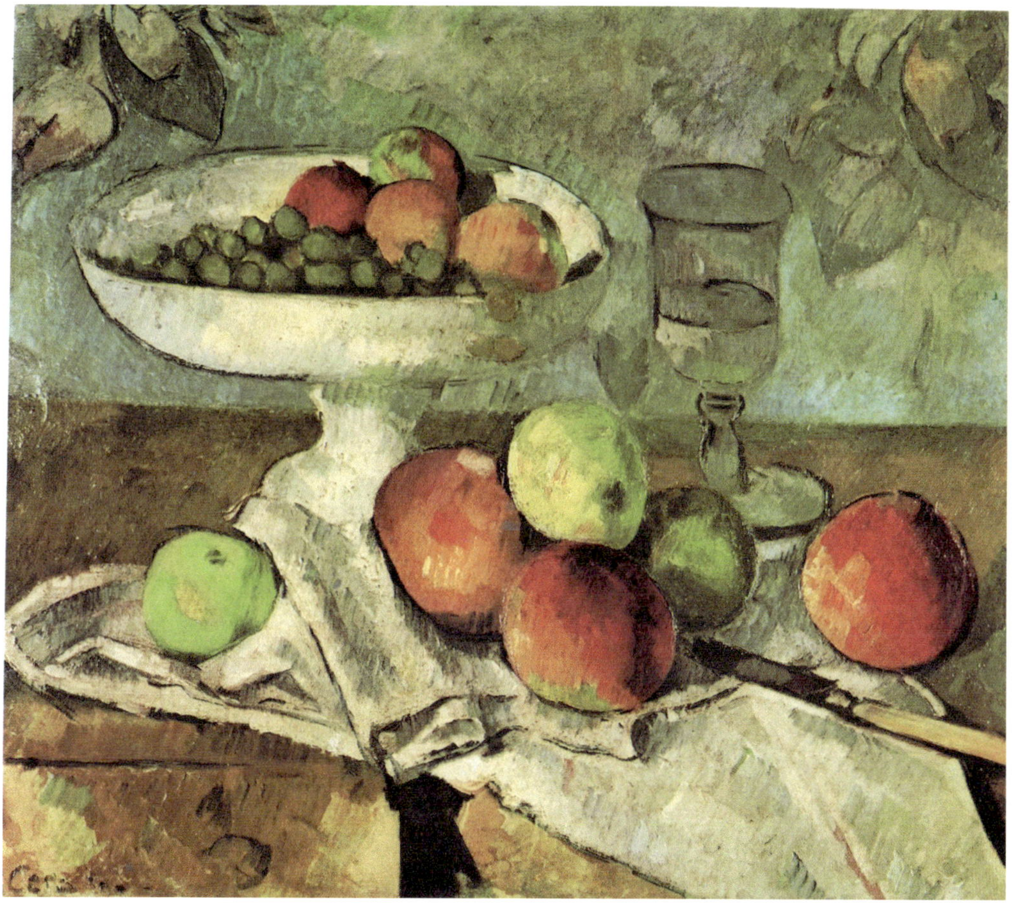

Figure 1. Paul Cezanne: Fruit-Bowl, Cloth, Glass and Apples (1879/80) Oil on Canvas, (46X55cm); Private Collection, Paris. (Some of the apples that conquered Paris and the world of art.)

The life story of the great artist, ***Paul Cezanne,*** is a demonstration of ***determination, endurance, persistence, ups and downs,*** and culminating in ***outstanding achievements.*** Paul Cezanne was born on 19 January 1839 into a well to do family in Aix–en-Provence, France. His father was an established Banker who was strict and authoritarian (Figure 2). The father cared mainly for business and making money. As the young Paul loved and practiced drawing and painting at his early years of childhood, those activities were not appreciated nor condoned by the father. Cezanne senior humiliated his son for wasting his time. He wanted his son to follow in the father's footsteps: study law, become a banker and make money. But that was not what the boy wanted. Fortunately for young Paul, it was ***his mother*** who believed in her son's artistic talent. She praised his work and she supported him when the father grew more and more worried about the boy's art work. Besides appreciating her son's work, she also kept Paul's early drawings in her room. This act enhanced the young boy's self confidence in art. The world at large should be most grateful for the mother's attitude and actions as they led to the birth of one of humanity's outstanding artists.

In 1849, Paul went to the Jesuit de Saint-Joseph till 1852, and next he joined the *College Bourbon*. At the College, he met his very close and intimate friend from Paris: ***Emile Zola***. Emile's father had earlier come to Aix to build a dam, but died during the construction work leaving his family in poverty. Cezanne's close friendship with Zola played a major role in his life. In 1858 Zola moved to Paris. However, their friendship continued through long, intellectual and affectionate letters. Zola was a very talented writer. He could see the writer, not only the painter, in Cezanne. He once wrote to Paul: ***"…you have a writer in you, just as I have. My verses are probably purer than yours, but yours are definitely more poetic and true. You write with your heart. I write with my head",*** (Zola to Cezanne, 1 August 1860). Though Cezanne did not pursue that talent, Zola's observation demonstrates that a ***true artist's talent is holistic***; a great painter does also feel the music and depth of words in the same way as he feels the harmony in his lines and colors.

In 1858, Paul bowed to his father's wishes and took up law studies at Aix University. Meanwhile, his father did not object to his son attending drawing classes in the evenings. Those classes were held at the Fine Art Museum in Aix. There Cezanne admired the museum's art work. After working as a banker, Paul realized that that was not his cup of tea. Thus with his mother's and sister's help he managed to convince his father to permit him to go to Paris to study art. In 1861, Paul abundant law studies and went to Paris where Zola awaited him. Unfortunately, Cezanne failed the Cole des Beaux-Arts entrance examination. And he returned to Aix to again work at his father's bank.

Yet, the artist in Paul Cezanne was certainly not happy as a banker and kept longing for art. In 1862 Paul again convinced his father that it is art that he loves, and the father yielded. Paul abundant the bank and devoted himself to art. With the father's help, he set a studio at Jas de Bouffan. In Paris, the father made him a monthly allowance of 150 francs. At Café Guerbois, Paul Cezanne met a number of, later to be world class painters, including: ***Camille Pissarro, Frederic Bazille, Claude Monet, Alfred Sisley, Pierre-Auguste Renoir*** and others. It was Pissarro who believed in and supported Cezanne in Paris. Cezanne the artist was thus born in Paris.

Nevertheless, his art was not appreciated by the academic art schools nor by many a classis art critique. Yet, Cezanne's new approach to art in the end conquered the world of art. He started a new school: ***Post-impressionism***. While he liked the work of the impressionists, he strongly felt that their work needed more structure and more depth. Those dimensions he masterly added. Cezanne excelled in landscape, portraits (figures 2 and 4), and still lifes (figures 1 and 3).

To appreciate Cezanne's school of art, the best way is to ***listen*** to the master himself. Here are some ***quotations from Cezanne*** on painting:

- ***"Let us read nature. Let us express our impressions and experiences in an aesthetic both personal and traditional. The strongest will be the one who touches the greatest depth and renders things entire – like the great Venetians."***
- ***"There is neither line nor modeling, there is only contrast. Once the colors are at their richest, the form will be at its fullest."***
- ***"I want to make impressionism into something as solid and permanent as the art in museums."***
- ***"There are no schools a priori. What counts is the art that is produced, as such. A picture is either good or bad."***
- ***"Two things matter in a painter: his eye and his mind. The two must give mutual support. Both need to be trained: the eye by studying nature, the mind by an orderly, logical approach to impressions and experiences. They create the means of expression."***
- ***"When he is confronted with the great masters, an artist learns to think. But when he confronts nature, he learns to see."***

- ***"I wanted to copy nature. And I failed. But I was content when I realized that the sun cannot simply be reproduced but must rather be expressed by some other means ... by color."***
- ***"My method, my law-book is realism. But a realism full of unconscious grandeur. The heroism of the real."***
- ***"To paint a picture means making a composition ...The test qualification for fine artistic conception is greatness of character."***

It is also important to appreciate how the world known artists perceived Cezanne. Here are some ***quotations by leading artists on Cezanne:***

- ***"However does he do it? The moment he dabs two touches of paint on a canvas, the result is brilliant." Pierre-Auguste Renoir.***
- ***"Our Cezanne is giving us all hope. I have seen, and have in my possession, a painting of remarkable power and energy. If (as I hope) he stays in Auvers for some time, and settles as he plans, he is going to astonish an artist or two who have been over-eager to condemn him." Camille Pissarro.***
- ***"It is a great misfortunate that this man has not enjoyed better support in his life. He is a true artist, but he has been brought to the point where he doubts himself far too much." Claude Monet.***
- ***"Cezanne was like the father of us all". Matisse.***
- ***"My one and only master." Picasso.***

In 1906, Cezanne passed away at the age of 67. His heritage is a remarkable work of art. His works are now exhibited in almost all the famous world museums. ***Paul Cezanne has, indeed, conquered Paris and the entire world of art, not only by an apple, but also by outstanding landscapes, portraits, and various versions of still life.***

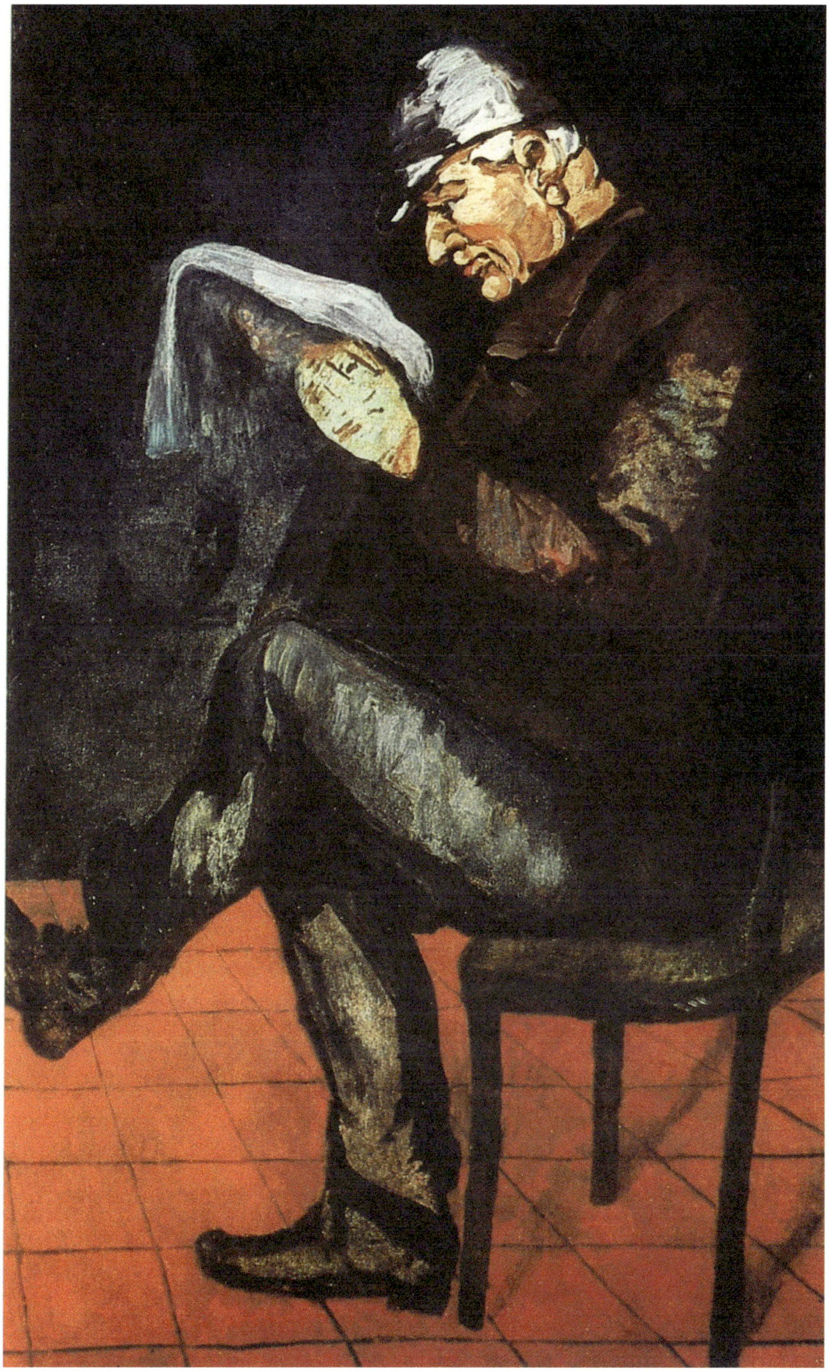

Figure 2. Paul Cezanne: The Artist's Father: Louis-Auguste Cezanne (1862) Oil on canvas (168X114cm); The National Gallery, London.

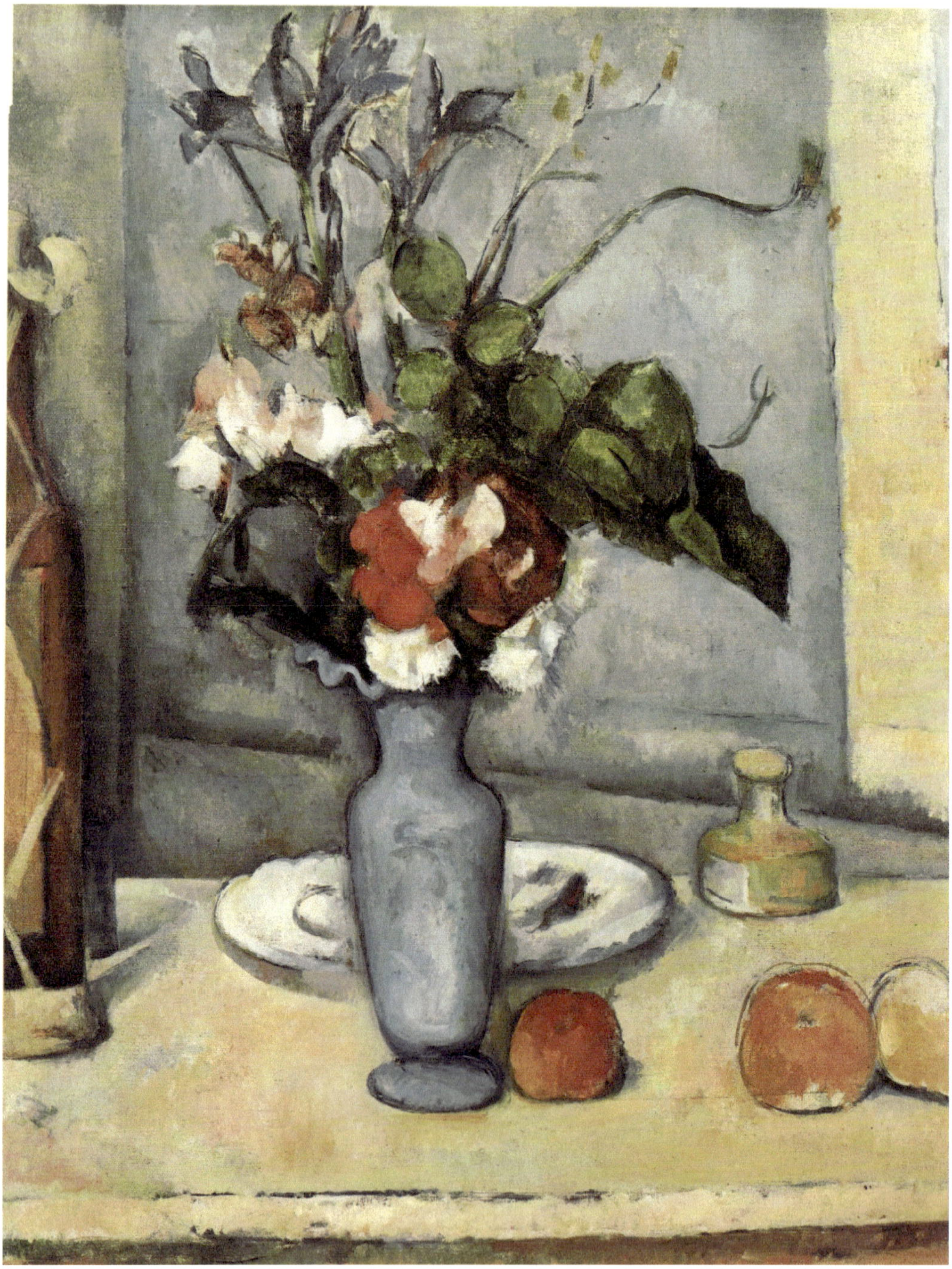

Figure 3. Paul Cezanne: Blue Vase (1885/87); Oil on Canvas (61X50cm); Musee D'Orsay, Paris

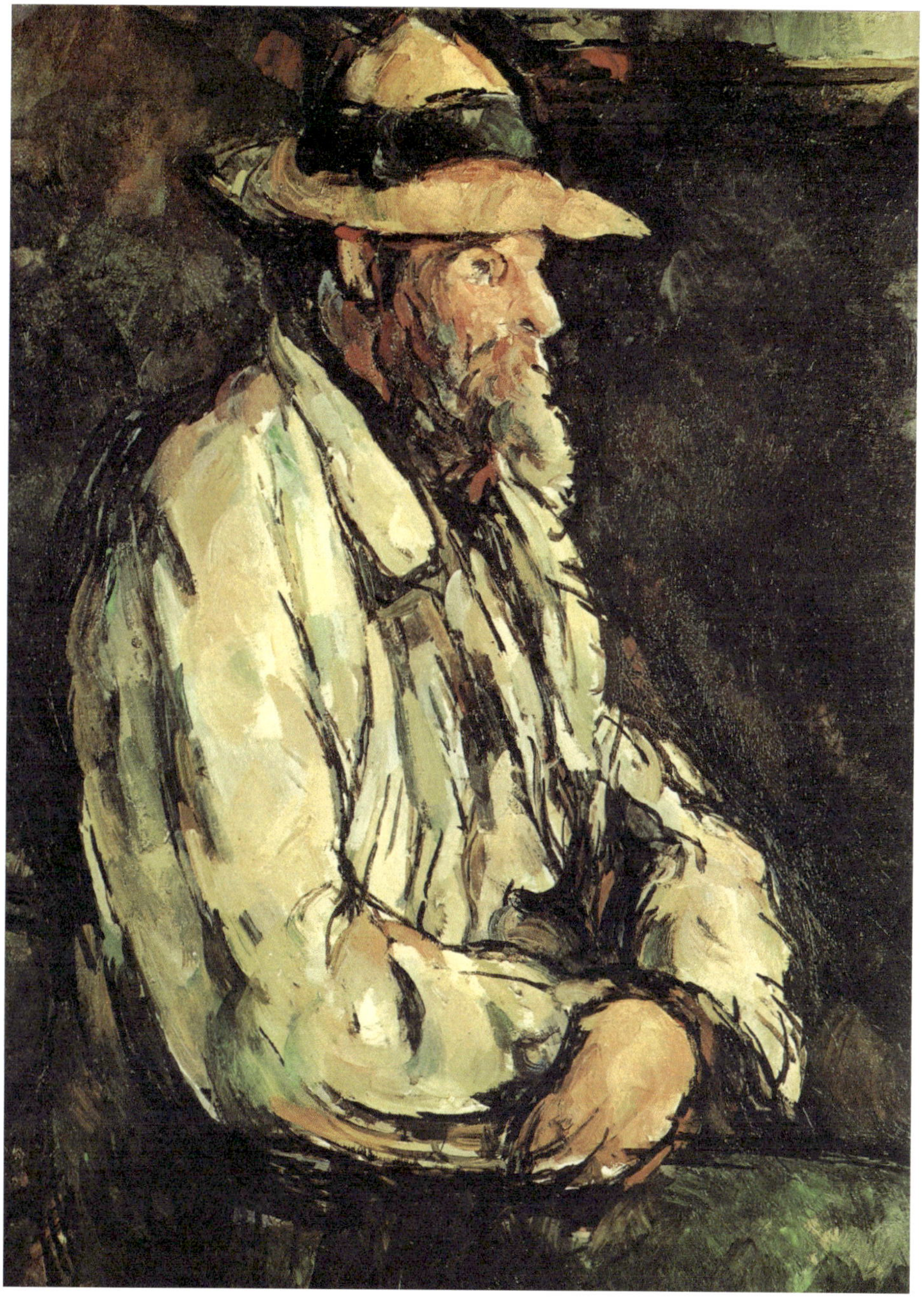

Figure 4. Paul Cezanne: Portrait of Vallier the Gardner (1906). Oil on Canvas (66X54cm). Private Collection, Switzerland.

"This is the very last painting by Cezanne. He completed it while on the death bed. The Artist passed away in 1906 at the age of 67. His outstanding art heritage illuminates most famous world museums in the West as well as in the east.

I.6. Roma Sparita

Ettore Roesler Franz's "Portraits" of the Eternal City

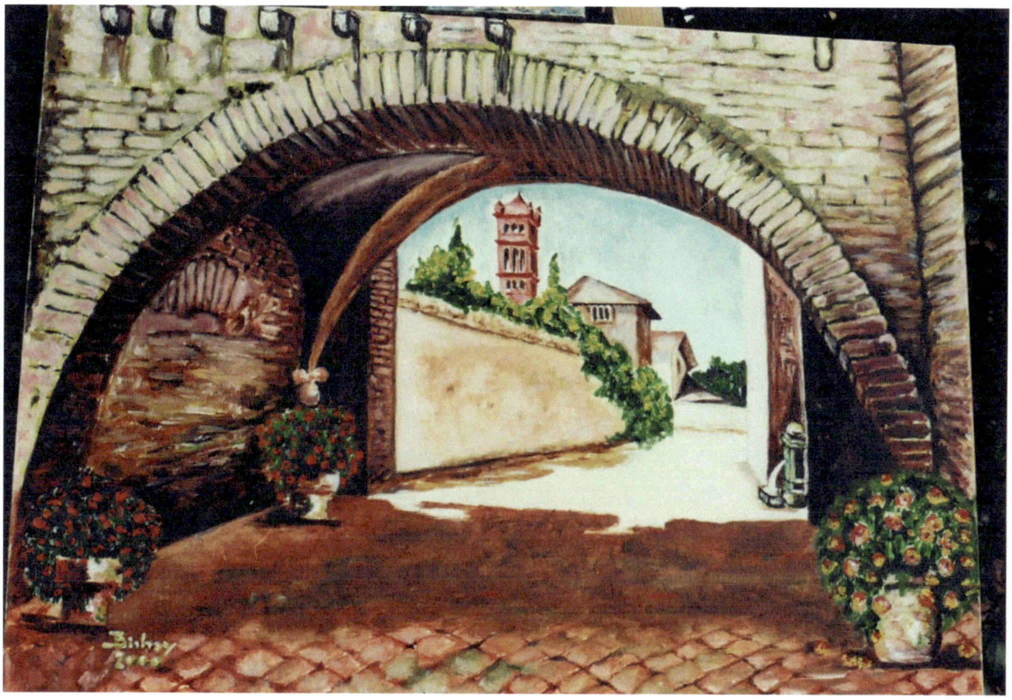

Figure 1. "Spring in Roma Sparita (1)". A painting by the author.
(Oil on Canvas 50X70 cm.)
(A Painting inspired by Franz's Roma Sparita (1845-1907).

Roma: the Eternal City has the power to absorb and assimilate those who live in it and ***love it,*** irrespective of their origin, ethnic background, sex, or religion orientation. One such example is ***Ettore Roesler Franz:*** A German Artist who was born in Rome of German parents in 1845. He loved Rome and was absorbed in the Eternal City's magic and beauty. His outstanding work of beautiful portraits of "Roma Sparita", (vanished Rome) is a testament of his love of the City. And the Eternal City's appreciation of his sentimental work was culminated in granting him, by acclamation, ***an honorary citizen of Tivoli.***

Like, Paul Cézanne, Ettore R. Franz started his career as a Banker in Rome. Yet the "Artist" in him revolted. He loved art. He loved old Rome. He excelled in water coloring. Franz was very much fascinated and actually fell in love with "Roma Sparita". He thus produced a number of outstanding ***water color "portraits"*** for old Rome. His paintings of Roma Sparita can, indeed, be considered as a kind of beautiful ***"portraits"*** of the Eternal City. Yes, those paintings are very much the same like ***"portraits of faces of distinguished human beings".*** For Franz's paintings of Roma Sparita do portray the inner feelings, the character and indeed the personality of old Rome. Such features can be perceived in ***the details; the city majestic arches; the dim colors; the movements; the donkeys; the donkeys' drawn carriages; the Cloths hung from the windows in the alleys; the balconeys; the blue collar workers and their ancient instruments; and the street water-fountains in Roma Sparita*** (Figures 3 and 4). Such detailed, painstakingly and very skillfully painted features in Franz's paintings of Roma Sparita do indeed render those paintings as "portraits" of the Eternal City. Sincerity, dedication and perseverance in carrying out his work are very clearly evident in Franz's "portraits" of Roma Sparita. He used to say: ***"Sincerity makes an artist great".*** How true.

The author has been fascinated by Franz's Roma Sparita. And thus inspired by Franz's outstanding work, he produced some paintings of Roma Sparita, (for example Figures 1 and 2). While inspired and guided by Franz's Roma Sparita, the author's paintings have two major differences from Franz's work: First, the author's paintings are ***in oil,*** using thick paint applied with both brushes and knives; while Franz's work was in watercolors. Second, the author's paintings are more colorful with a lot of ***added flours and blue skies*** characteristics of Rome. The author's work could, thus, be labeled: ***"spring in Roma Sparita"*** (Figures 1 and 2).

Ettore Roesler Franz was born in Rome on the 11th May, 1845 to a German family as was noted above. He completed his studies at the *Istituto dei Fratelli delle Scuole Cristiane,* the *Collegio di Propaganda Fide and the Accademia di San Luca,* where he studied ***architecture.*** Until he was 30, Franz was involved in his family's business affairs, in particular as a banker. However, from 1875 till the end of his life (1907), he spent his time exclusively painting. He founded, along with *Nazareno Cipriani,* the Society of Watercolorists in Rome, of which he was president many times.

In 1883 he participated in the Palazzo of the Exhibition with his first series of watercolors entitled ***"The memory of a past age".*** Franz dedicated his great work to Rome: His beloved City. In 1885 and 1887 he exhibited several paintings at the London exhibition at the Royal Institute of Painters in Water Colors. In 1888 he exhibited in Berlin; in 1890 he exhibited at the Exhibitions in Dresden and Vienna, where his watercolors were awarded a ***gold medal.*** In 1902 he took part in the Italian Pavilion in Saint Petersburg. On the 6th of February 1903 he was granted, by acclamation, ***an honorary citizen of Tivoli***.

Franz owed his fame to his fabulous series of watercolors of Roma Sparita. He, however, portrayed other sites such as Tivoli and its surroundings, the Villa d'Este, the Villa Adriana, the Albani Hills, the Via Appia, Ninfa and the Vatican marshes.

On 26th, March 1907, ***Ettore Roesler Franz*** passed away in his adopted and beloved home: Rome, at the age of 62.

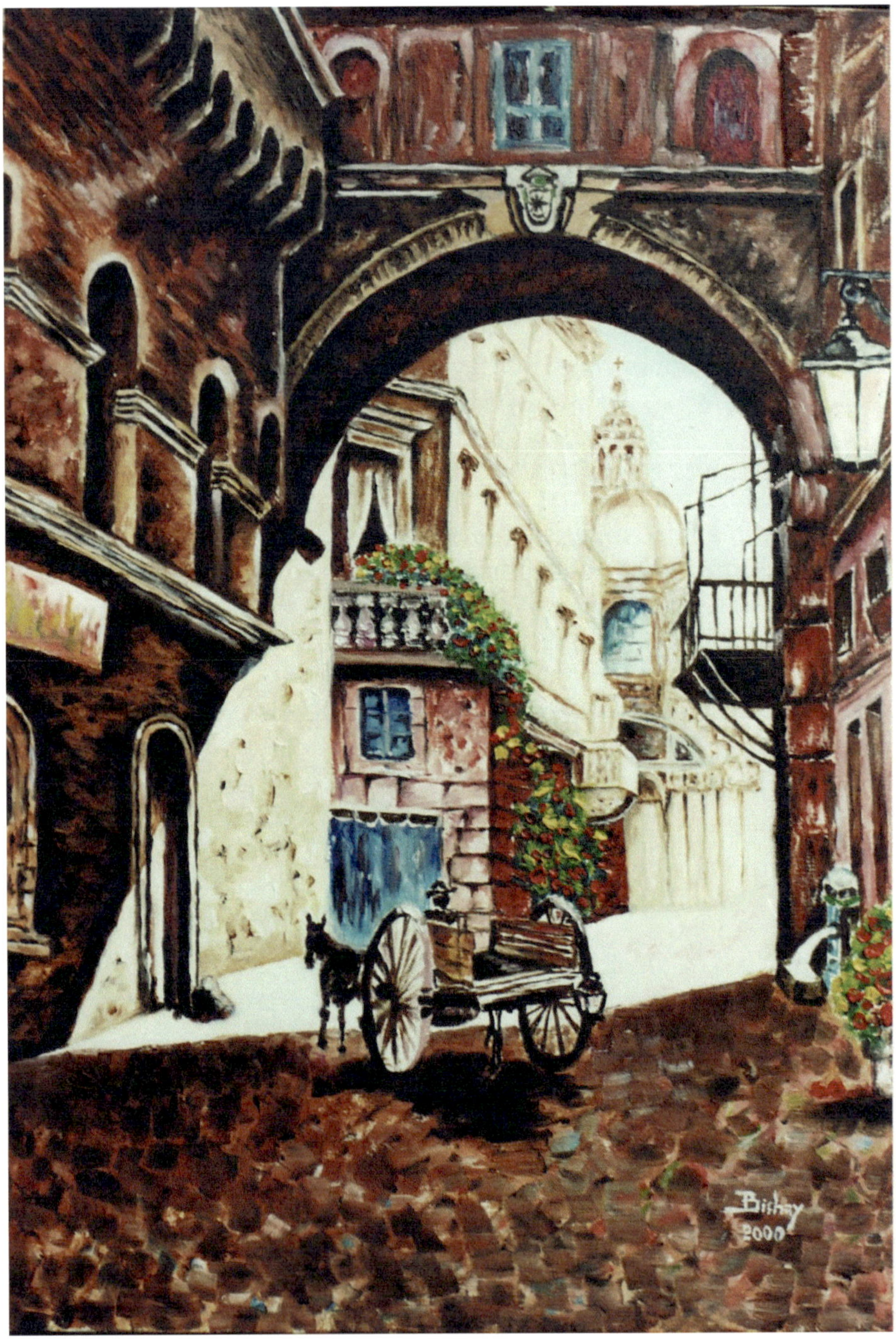

Figure 2. Spring in Roma Sparita (2);
A Painting by the author inspired by Franz's Roma Sparita (1845-1907).
(Oil on Canvas 70X50 cm).

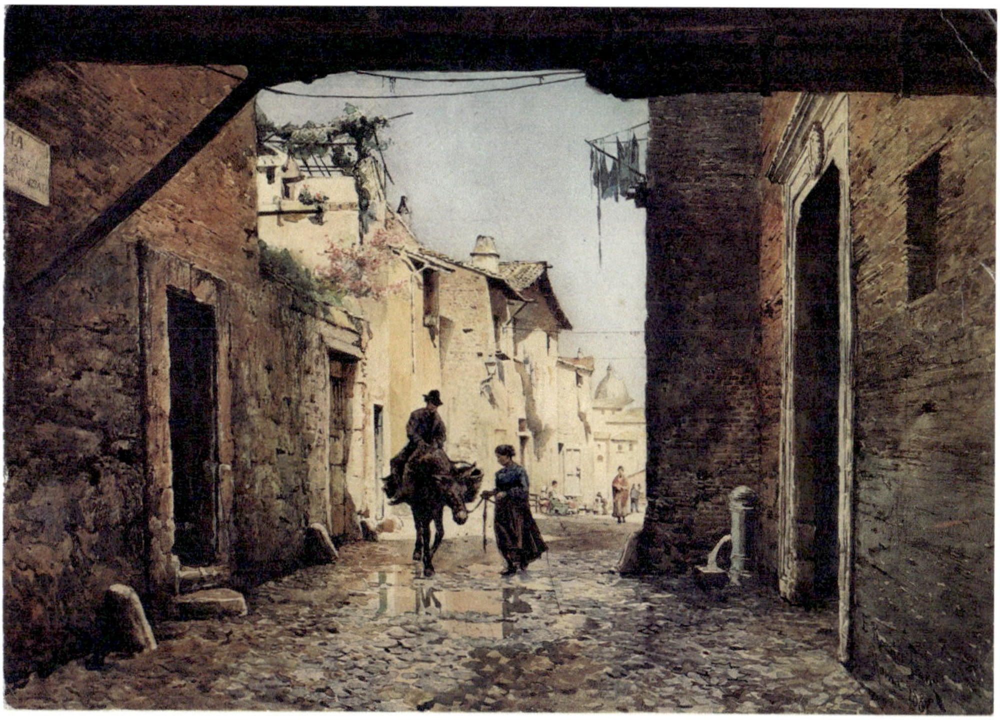

Figure 3. Roma Sparita (1) by Ettore Roesler Franz (1845-1907): Museo di Roma.

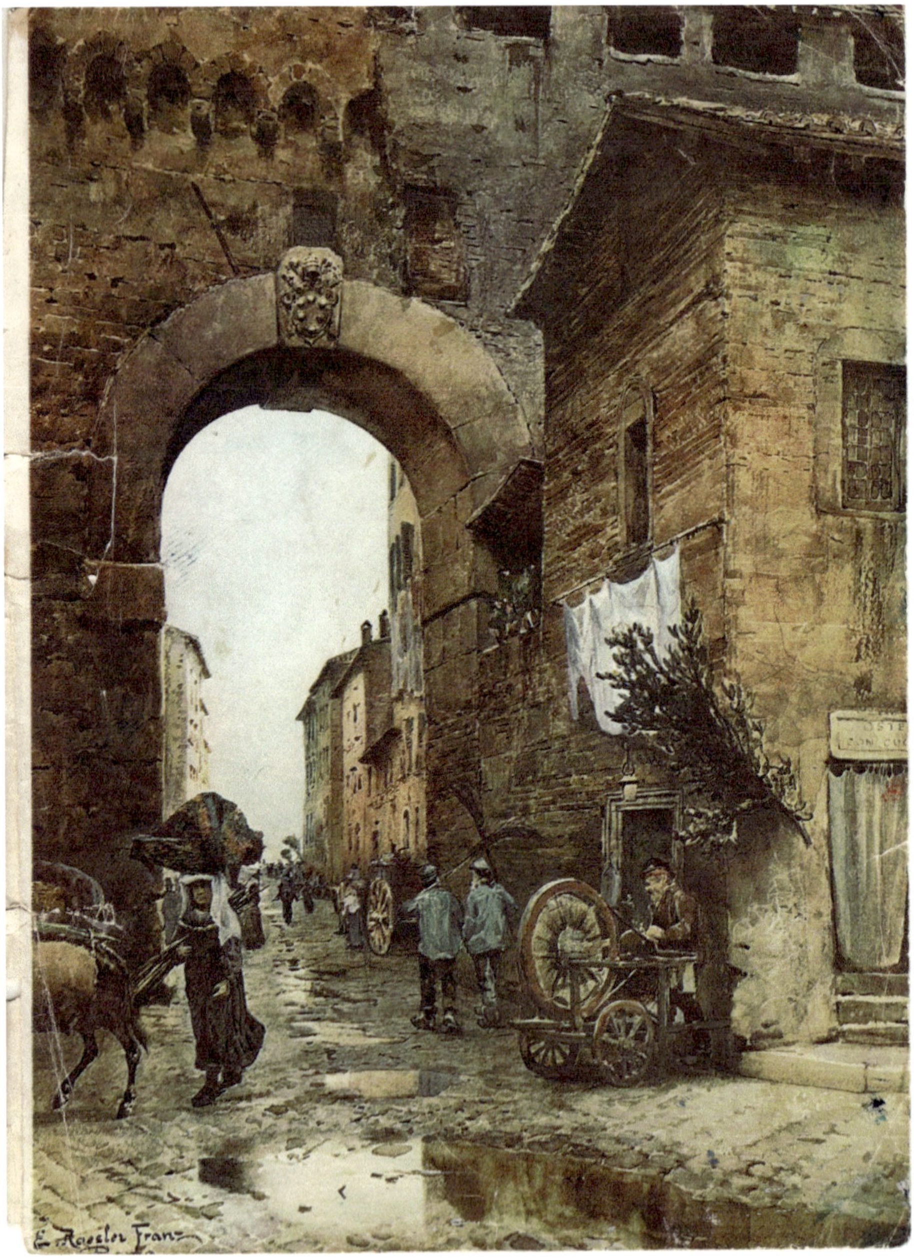

Figure 4. Roma Sparita (2) by Ettore Roesler Franz (1845-1907): Museo di Roma.

I.7. Towards a Holistic Approach to Oil Painting

Based on my experience as an amateur painter and drawing on selected literature on the subject, I found that a number of key principles, constituting an effective holistic approach to oil painting, can be very helpful in realizing the painter's objectives. This approach is highlighted in this article

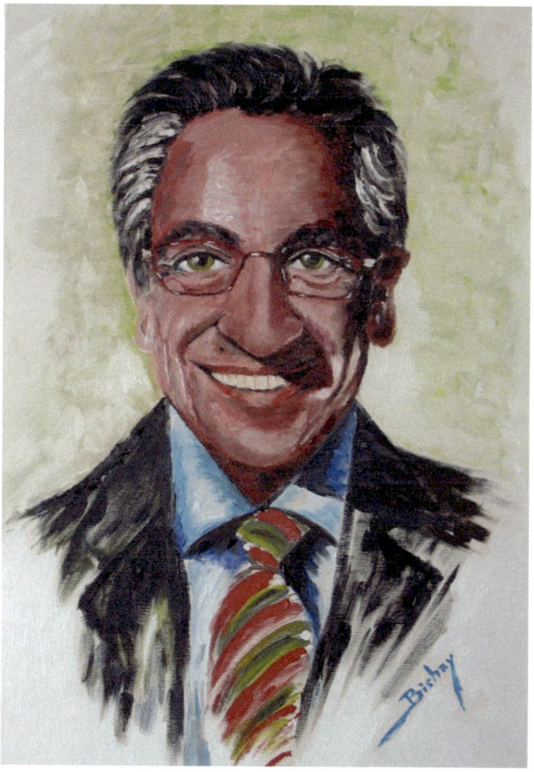

Figure (1). "Smiling": A self-portrait by the author (oil on canvas: 60X40 cm).

Introduction

In the literature on art techniques for oil painting, there are so many theories, principles, approaches and a large number of schools of painting. For a professional painter, he/she has, indeed, to study, analyze and reflect on that very rich literature. However, for the amateur painter I have singled out in this article only a ***few*** techniques that, in my judgment and experience, ***are simple yet effective*** elements that would help him/her to achieve his/her objectives of painting. Those elements constitute a holistic approach to oil painting.

Notwithstanding the above, we must keep in mind that art (all sorts) by definition has a significant element of ***"subjectivity".*** And that is what makes it innovative, dynamic, interesting and most enjoyable. Note that we say "an element of" and not "totally subjective" for there are a number of basic "objective" principles that can help the artist to achieve his/her objective. Furthermore, in view of that subjectivity element in the art of painting, the selection of and the emphasis placed upon each one of the key "objective" painting principles vary from one painter to another, one subject to another and one media to another.

With the above in mind, the following are my ***selected*** elements of a holistic approach to oil painting.

Towards a Holistic Approach to Oil Painting: An Integrative Process

I wish to postulate that there is a holistic approach to painting. That is, the art of painting is a process, a holistic process, not an event. Although the discussions to follow on the selected painting techniques are presented in sets of separate elements (to facilitate presentation), the effective approach is indeed a ***holistic approach***. Thus, at the implementation phase of painting the artist ***has*** to fully ***integrate*** those aspects in a single holistic approach in order to realize what he/she aims to achieve from his/her work.

Key Components of a Holistic Approach to Oil Painting

- ***Focal Point: The Concept and the plan***

In one of my earlier articles (Section I.3.), I strongly argued that ***before*** you start your painting, you should have a ***concept*** or an ***objective*** of the envisaged painting. The painting's objective culminates in your ***focal point***. You should next have a ***plan*** for your painting to achieve that objective. A painting plan is how the painting will be implemented and how it will be read by the viewers. The integrated objective, focal point and the associated plan demonstrate the artist's way of seeing things. In the above-mentioned article, a number of examples of paintings' objective, focal points and plans were highlighted.

Other ***examples*** are presented here. For instance, in Figure (1) above, the objective of the painting was to show the happy smile in that portrait. The focal point was, thus, the smiling face. The plan of the painting was to emphasize the light movements on the face's muscles as the subject was smiling. The eyes were cast in a happy mood. Using happy and bright colors both in the face and the clothing helped to bring the radiant in the smile.

- ***Composition: Guiding the Viewer's Eyes***

Once you have decided on your painting's concept or objective, the focal point and the execution plan, you would have a ***mental image*** of your painting. Next, you compose your painting to realize that image. Although the phrase "painting composition" could be intimidating, it should not. When, you have a mental image of your painting, composition of the painting becomes easy and simple. In composition you have to mainly focus on the following three elements: ***direction, balance and integration***.

Under ***"direction"*** you design your painting such that you guide the viewer's eyes towards your focal point. In general the focal point of your painting should ***not*** be in the ***centre*** of your painting for that could make the painting too symmetrical and boring. The components of the painting should, in a subtle way, guide the viewer's eyes to your focal point. For instance in Figure (2), the focal point is the water coming from the pot into the girl's mouth. The main elements leading the viewer's eyes are: The movements in the girl's face, eye, neck and shoulders. The position of the water pots and the girl's left arm further direct the viewer's eyes to the focal point.

Under ***"balance"*** you balance the "weights" of various shapes in your painting. That is because in painting, shapes are conceived as if they have weights: A shape may look "heavy" if it is large, dark, busy or very bright. You have to aim at balancing those imaginary weights in your painting. This element in particular is especially subjective. Thus balancing the weights in a painting composition is based on how you ***feel*** about those imaginary weights (see Figure 2).

Under ***"integration"*** you avoid a painting of scattered components. Shapes or major components of the painting should lead to each other so that you should have a flow to the painting structure. This aspect is especially important in the composition of ***"still life and flowers"***, (see Figure 3).

- ***Color and Value: From Two to Three Dimensions***

Of the many theories of ***colors,*** I found that a few color principles work very well. Those are: First, ***contrast cool*** colors (e.g., blue, light blue, white...) with ***warm*** colors (e.g., red, brown, orange, yellow…). Second, contrast neutral against pure and intense colors. Third, repeat colors intelligently around the painting until it looks virtually balanced.

The word ***value*** refers to "lightness" or "darkness" of a particular color in the painting. Value is very important to reflect ***lights and shadows***, and, thus, value can create the illusion of transforming the flat two dimensions image on the canvas into a seemingly three dimensions painting. This is achieved by making objects and spaces seem to ***"recede" (darker) or "come forward" (lighter).***

When a Painting is finished?

Finally you have to know when your painting is finished; otherwise you might reach an overworking stage in your painting. An overworked painting loses its freshness and becomes dull. The answer to that question was presented in my earlier article referred to above under the title: "When a painting is finished?" (Section I.3.). As noted in that article, when your perceived objective and the essence of the plan of your painting are achieved, you stop. Your painting is finished. This approach makes the painting process thoughtful and well guided.

Concluding Remarks

While the above proposed holistic approach will certainly help you in going about your oil painting in an effective and working manner, I wish to reiterate the above-mentioned innovative subjectivity element in painting. Thus, while the proposed holistic approach will be helpful, ***you have to use your own feelings, your own personal perceptions and your own taste in your painting.*** Only then you will be ***innovative,*** and you will ***reflect your own personality in your art work***: ***A process that is most enjoyable and highly rewarding.***

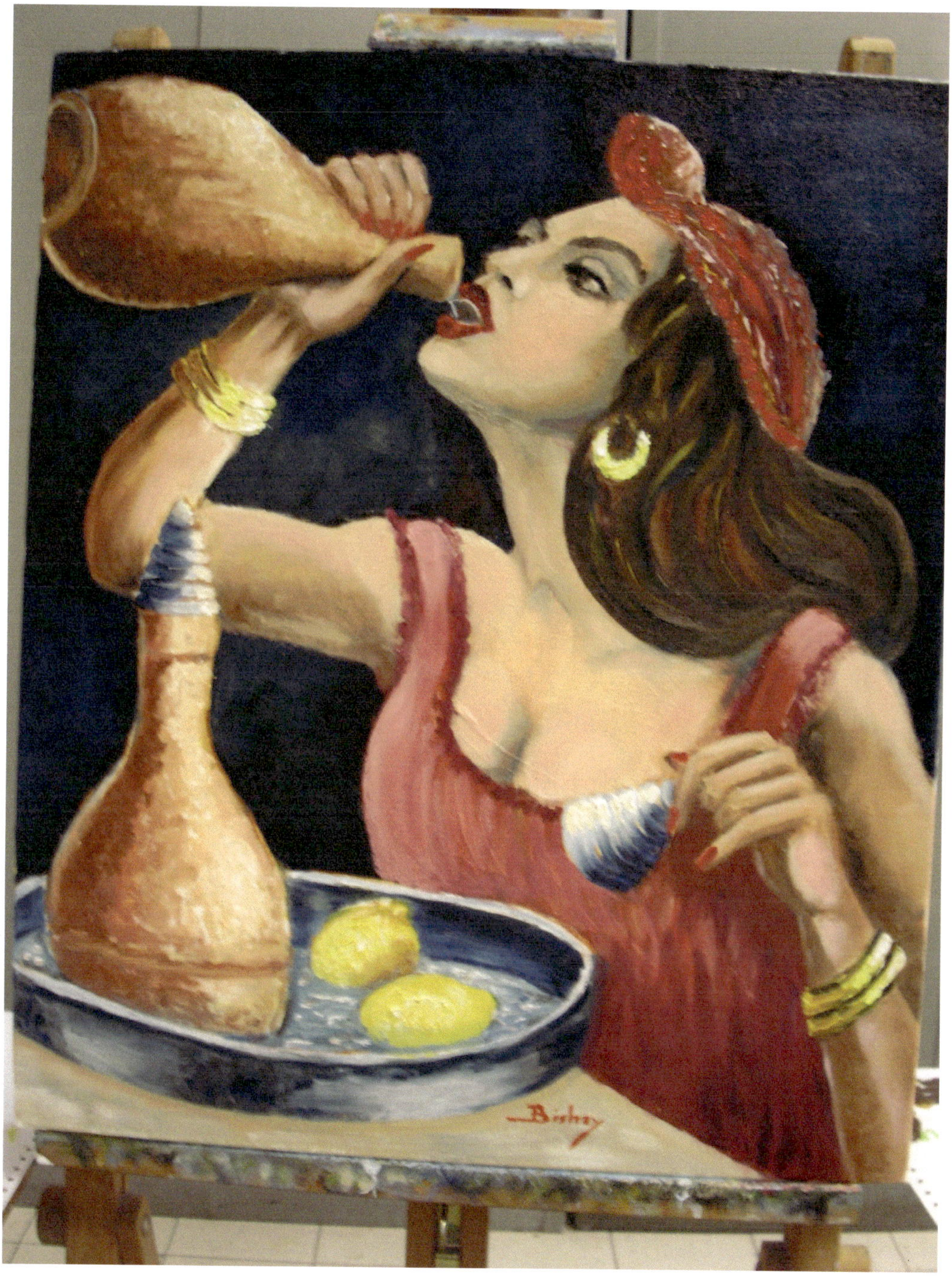

Figure 2. "Thirsty": A painting by the author (oil on canvas: 50X40 cm)

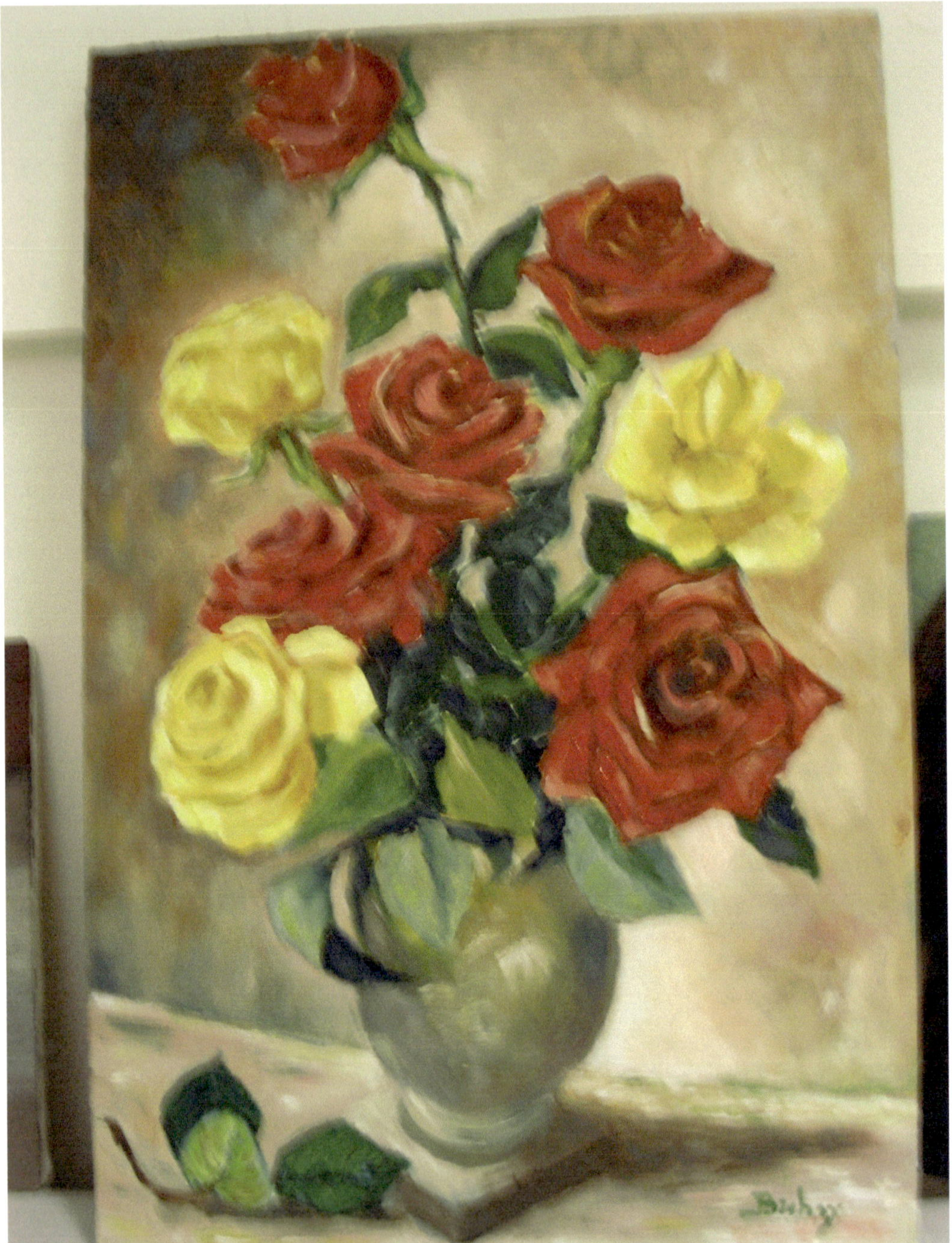

Figure 3. "A Portrait of Roses": A painting by the author (oil on canvas: 60X40 cm)

I.8. Italian Mosaic:

A Fascinating Art

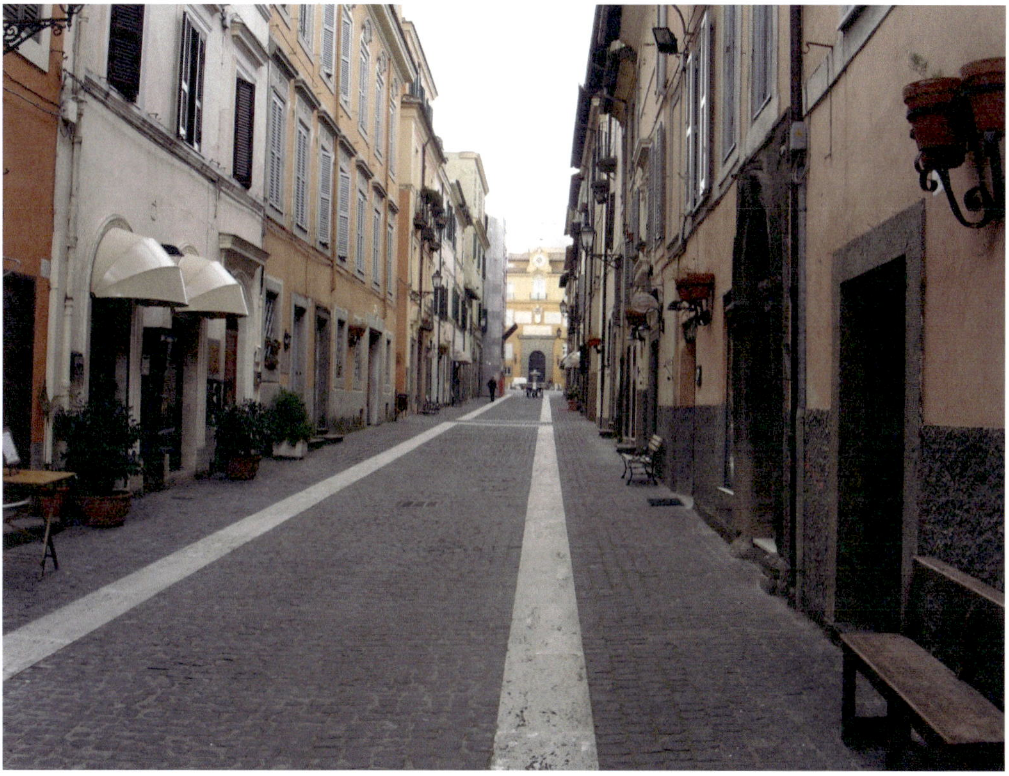

Figure 1. Castelle Gandolfo where Artisan Mosaic workshops existed in Rome.

Introduction

The Italian mosaic is a fascinating form of art. In this art, small tiles or fragments of marble stones are used to create patterns or pictures. The final products represent an appealing marriage of intricate colorful patterns and images reflected in the austerity and solidity of the marble stone. The marble mosaic work is almost alive: it grows in color and reflection in the same way as marble does; and it could last for centuries. The Italian marble mosaic tradition has existed for nearly four thousand years, with origins mostly believed to have been in the Greek and the ancient Egyptian architecture. In ancient Egypt, there existed many decorations in glass mosaic in Pharaohs' palaces and temples. The Egyptians were the first to discover the fusion of glass, and to create a flourishing industry which eventually led to the decoration of their ships. It is likely that the usage of glass mosaics reached Italy from Egypt.

Mosaic art is frequently present in Roman dwellings; especially old dwellings. For example in those beautiful homes, great mosaic works can be found on walls, stairs, window sells and floors. The craft, however, has evolved through the ages from a decorative art into an independent, fascinating, colorful and very effective art as such. The amount of time and labor expended to create an outstanding mosaic piece are phenomenal, yet the beauty and durability of the final products justify the effort.

A mosaic coffee table by the author

A few years ago, I visited Petra and an artisan workshop for Jordanian mosaic, and the beautiful villa of my friend Mr. S. Sabanegh, which has outstanding Jordanian mosaic works, in Amman Jordan. I was very much impressed with that art, and fell in love with its fascinating color, texture and outstanding handcraft. Thus, upon my return to Rome, the city of art, I decided to study, learn and do some mosaic art work. As I investigated the matter, I learned that in Castel Gandolfo there are certain Artisan workshops where I could be a "student" for those Italian masters and would learn the trade. So I did. It took a few months to appreciate and learn this fascinating art. And here I summarize the steps to be followed to complete an Italian marble mosaic piece.

Step1: The Design

Naturally you start your "project" by deciding what do you went to produce. I decided to do the top of a coffee table (size: 80X40 cm.). Next you have to draw the picture to be done by the mosaic. So, I drew some lotus flowers by pencil on special kind of heavy paper. Of course the paper's size should be identical to the top of the table where the mosaic piece will rest for good.

Step2: Cutting and sticking the marble according to the design

Once you have the design on that special paper, you start the hard work of cutting and sticking the marble. The work at this step is, indeed, difficult and requires a lot of patience. You get the colored pieces of marble in squares of about one square inch each. Then you cut it with a "vicious" hammer that is sharp, heavy and especially designed for this job. (As you start learning this art, you might mange to hit and "hurt" your finger in your left hand, but with practice you will overcome this difficulty.) You cut the marble into much smaller pieces and colors according to the design. The more curves and colors there are in the design the more beautiful the work will be, but of course the more difficult and demanding. You stick these small pieces with special glue on the penciled drawn design. This task is done manually and very carefully. The size, shape, angles, and colors of the cut small pieces are very important in deciding how accurate you will be able to realize your original design. Depending upon the size of the final product, you may have to work on part or all of the paper with the design. In our example, I worked first on one half of the design, and later on the other half. Figure 2 shows the progress in that step in only one half of the design; and Figure 5 shows the completed design.

Step3. Preparing the cement base

Once the design has been realised by the marble pieces on the paper, you start preparing the cement for placing it in its final place (in our example the top of the coffee table). At this step a sufficient amount of cement is prepared with the appropriate thickness (See Figure 3).

Step 4. The cemented table top

In this step the cement is placed on the top of the table (see Figure 4). The surface should be very carefully finished and completely horizontal.

Step 5. Placing the piece on the top of the table: the final product

This is very delicate and difficult step. Here the paper with the mosaic design will be placed ***upside down*** on the cemented top of the table. After ensuring that the piece is perfectly in place, (with the paper on the top), using special chemicals, you very carefully eliminate the paper. This process will be done very slowly, and centimeter by centimeter. Eliminating the paper could thus take several hours. You also have to be very careful in ensuring that this process will not affect the levelness of the surface. Once all the paper is out, you start another exercise of cleaning the top. This leaves the final picture clean, bright, and leveled. The work at this point is completed, and you would finally need to leave it for about a week to completely dry up. Figure 5 shows the final product. And Figure 6 shows the mosaic coffee table at the author's home.

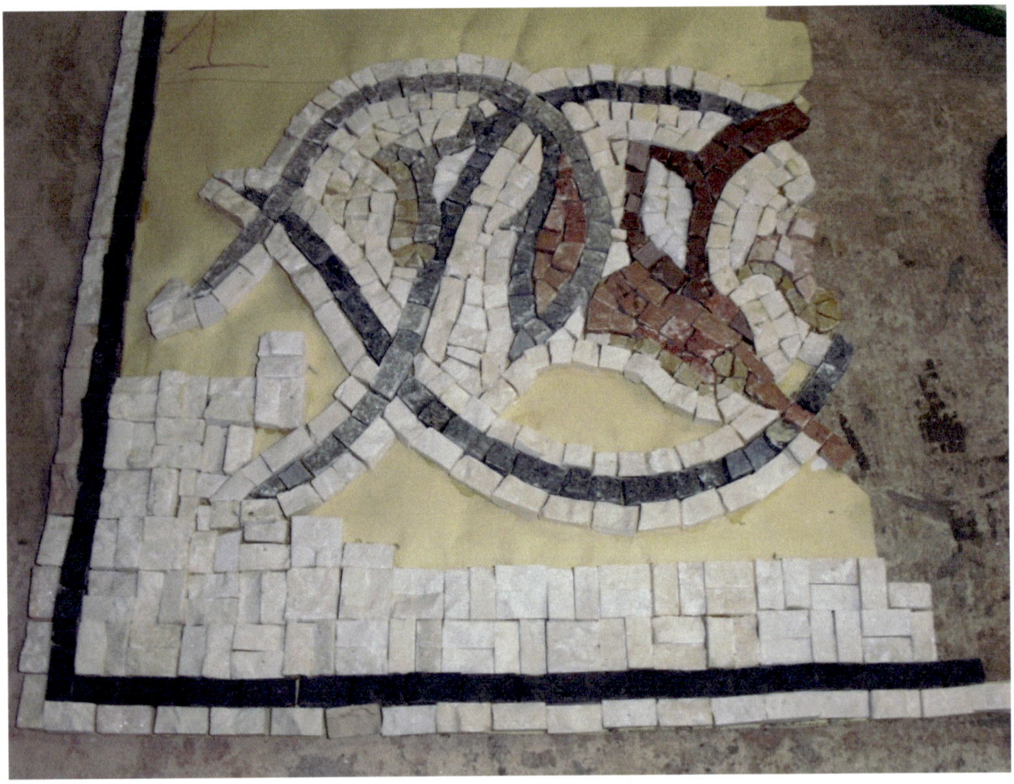

Figure 2. The design on special paper and starting to cut and paste some marble pieces.

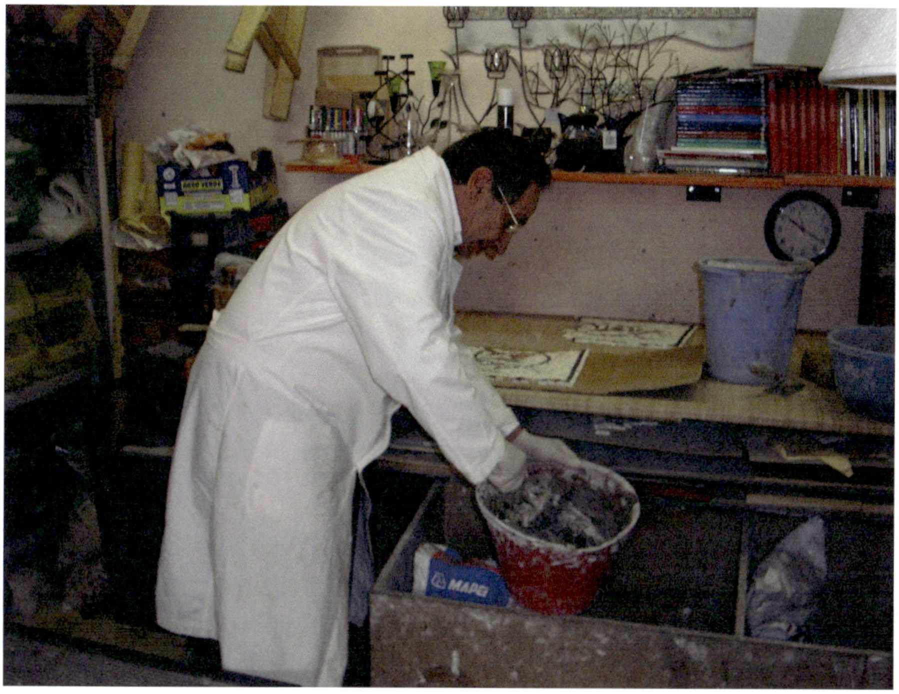

Figure 3. The author preparing the cement base as the foundation for the mosaic piece to be placed on the top of the table.

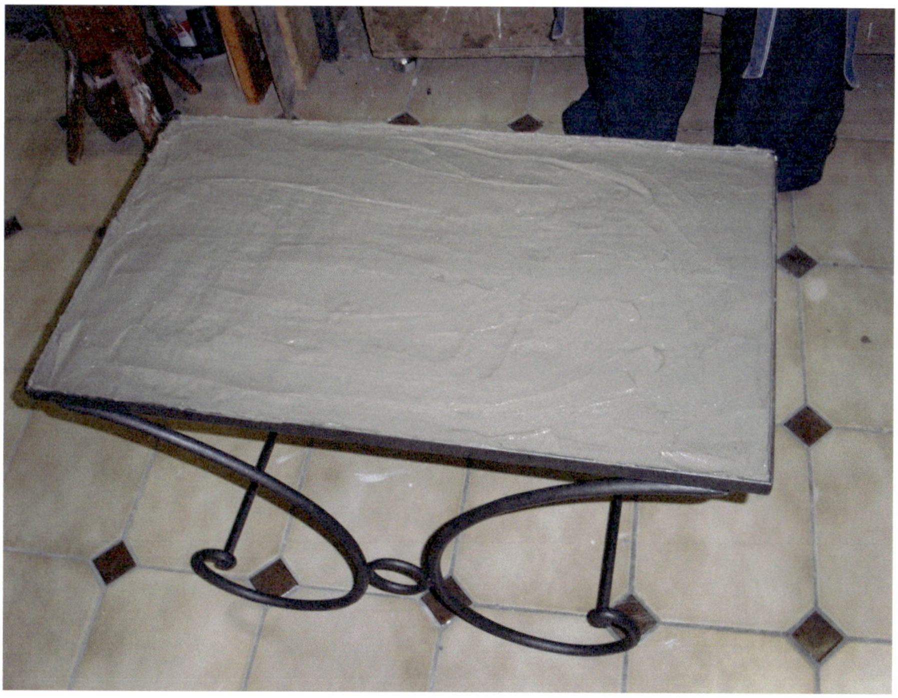

Figure 4. The cemented top of the table prior to placing the mosaic piece.

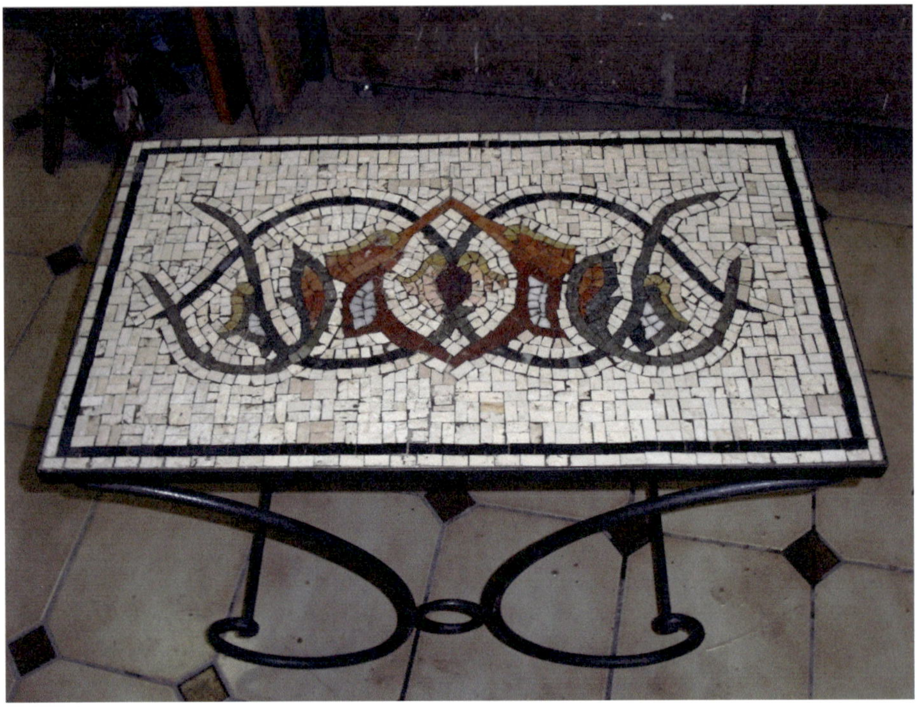

Figure 5. The final product: A mosaic coffee table.

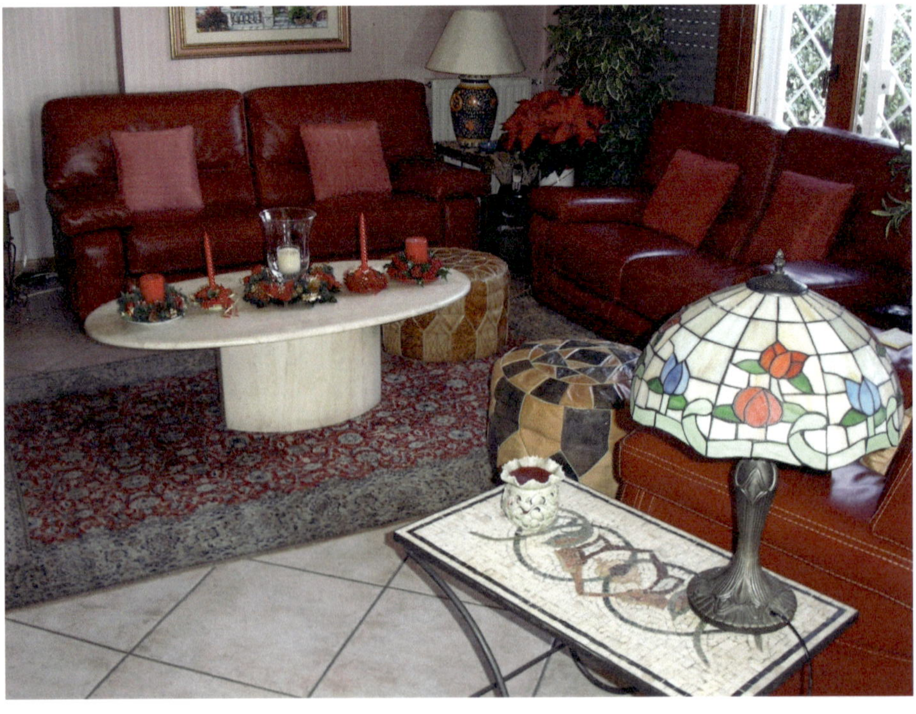

Figure 6. The mosaic coffee table at the author's home.

I.9. Spring in Rome

The season of growth and flowering

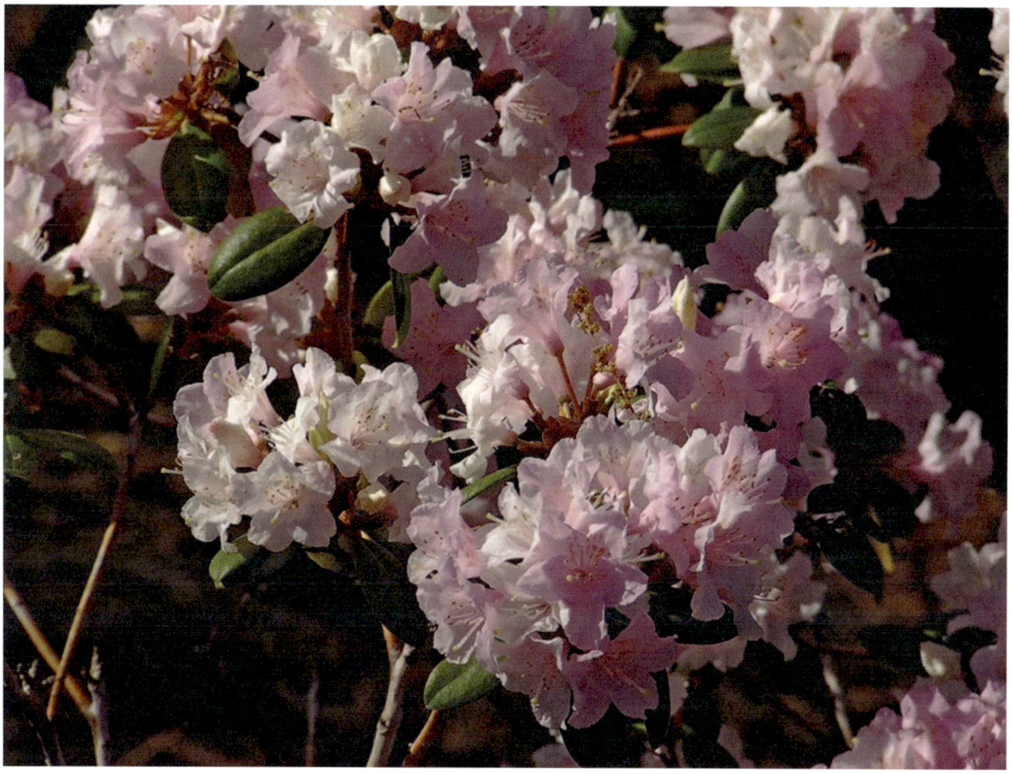

Figure, 1. Anatomy of the flower Rhododendron* 'Manitau' *Ericaceae: A photo by the author of a common flower in spring in Rome.

Spring is the season of growth, renewal, and flowering of new life. Spring is the season for love and flowers. The term is also used more generally as a metaphor for the start of better times and times for emotional affection as in "spring fever".

Spring in Rome is especially beautiful. Outstanding colorful flowers are blooming everywhere. Lovely weather with warm and kind sun is enjoyed daily. And every day of the spring in Rome is, indeed, a feast.

One of the most beautiful streets in the spring in Rome is via "Ignazio Guidi" in Piazza Navegatori. In spring, the entire street becomes full of trees with the pink flowers: ***Rhododendron* 'Manitau' *Ericaceae*.** Figures 1 shows the anatomy of the flower. Figure 2 shows the "flowering" street in April every year. Figure 3 shows a close up of the flowers in that street.

When such beautiful flowers are perceived in relation to an old Roman setting, an outstanding scene emerges. Figure 4 shows an oil painting by the author where a perception of a number of very beautiful elements from the old and the new in spring in Rome is demonstrated. The painting is inspired by the spring in via Ignazio Guidi presented against an old Roman background.

Spring in Rome is indeed very beautiful.

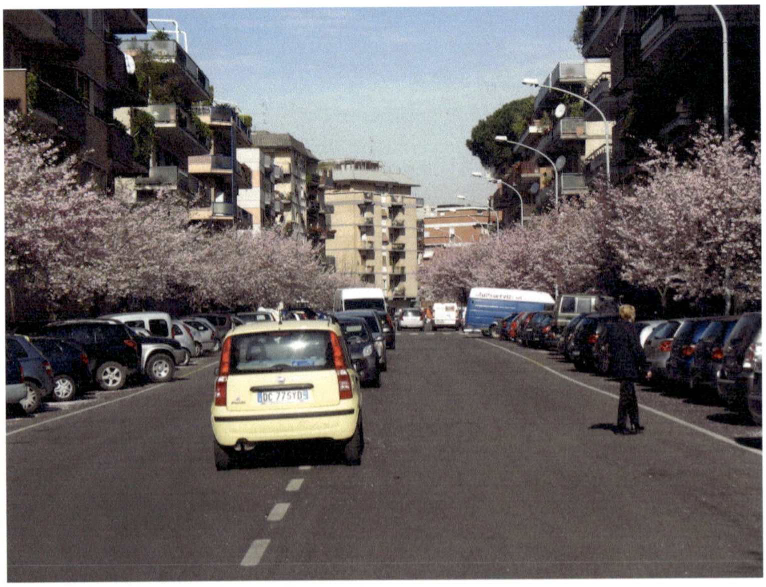

Figure 2. Via Ignazio Guidi in spring.

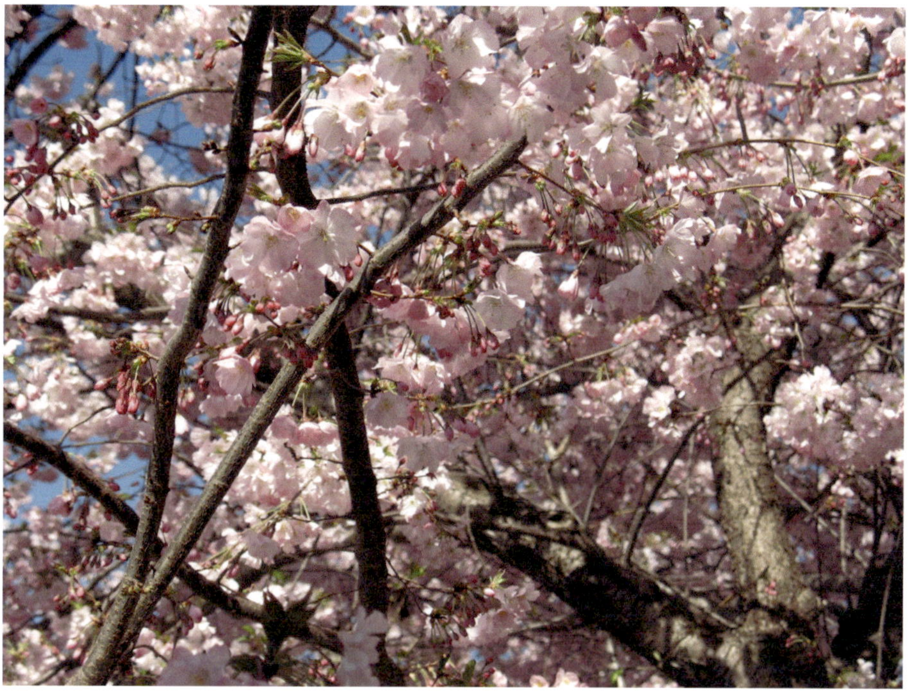

Figure 3. Close up of the flowers in via Ignazio Guidi.

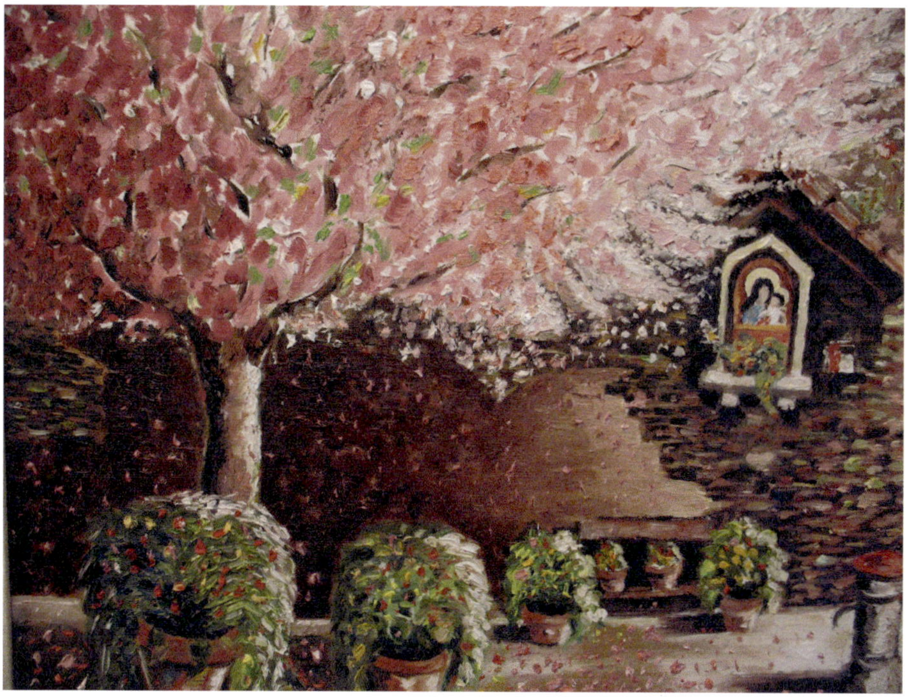

Figure 4. Spring in Rome: an oil painting by the author (Oil on canvas: 40X50 cm).

I.10. Autumn in Rome

When every leaf becomes a flower

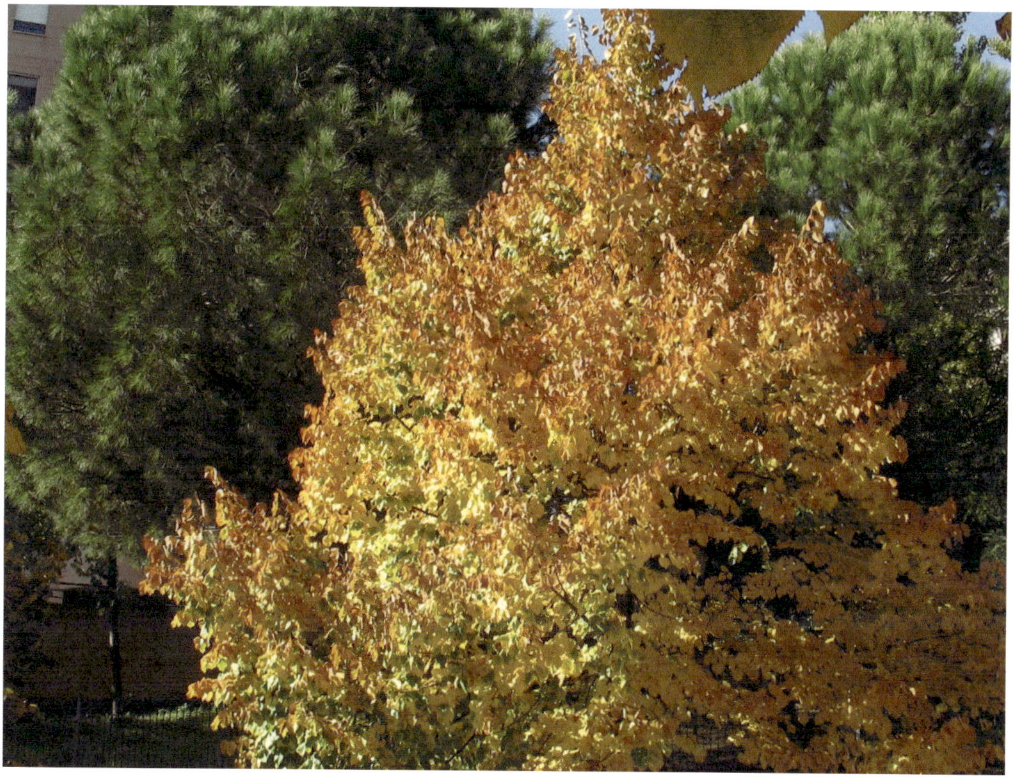

Figure 1. A photo by the author in Via Delle Costellazioni; Rome; November 2009.

Autumn is the season associated with the transition from warm to cold weather. It is the season when every leaf becomes a flower. It is also the season of ***primary harvest.*** The season is further associated with nature beauty: Golden falling leaves from the trees. Most cultures enjoy ***autumn celebrations of the harvest***, as an important season on their calendars. For instance, the US celebrates ***Thanksgiving*** in mid-autumn.

The word ***autumn*** comes from the Old French word *autompne* (*automne* in modern French), and was later normalized to the original Latin word *autumnus* There are rare examples of its use as early as the 12th century, but it became common by the 16th century.

Before the 16th century, ***harvest*** was the term usually used to refer to that particular season. However, with the significant rural-urban migration in the developed countries, the word ***harvest*** lost its reference to the time of year and came to refer only to the actual activity of reaping the fruits of the land. And ***autumn*** (as well as ***fall***) replaced it as a reference to the season between spring and winter.

The word ***fall*** is now mostly a ***North American English*** word for the season. It traces its origins to the old ***Germanic languages.*** During the 17th century, English immigration to the colonies in ***North America*** was at its peak, and the new settlers took their language with them. While the term ***fall*** gradually became obsolescent in Britain, it became the more common term in North America.

Autumn in Rome is beautiful as can be seen in the attached photos by the author of the season in selected streets in the city.

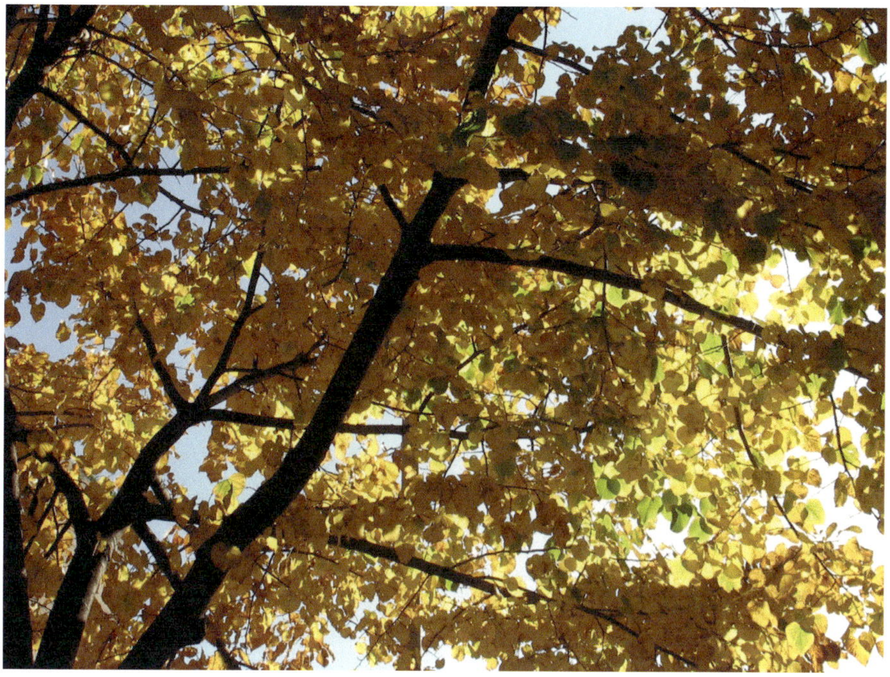

Figure 2. A photo by the Author of autumn in Via Degli Astri, Rome, (November 2009).

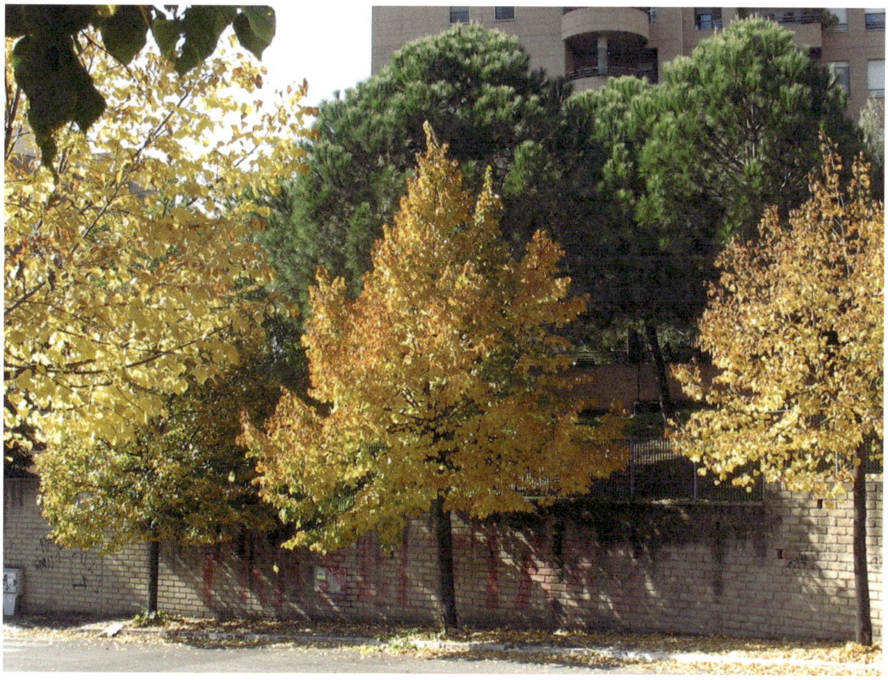

Figure 3. A photo by the Author of autumn in Via Caterina Troiani, Rome, (November 2009).

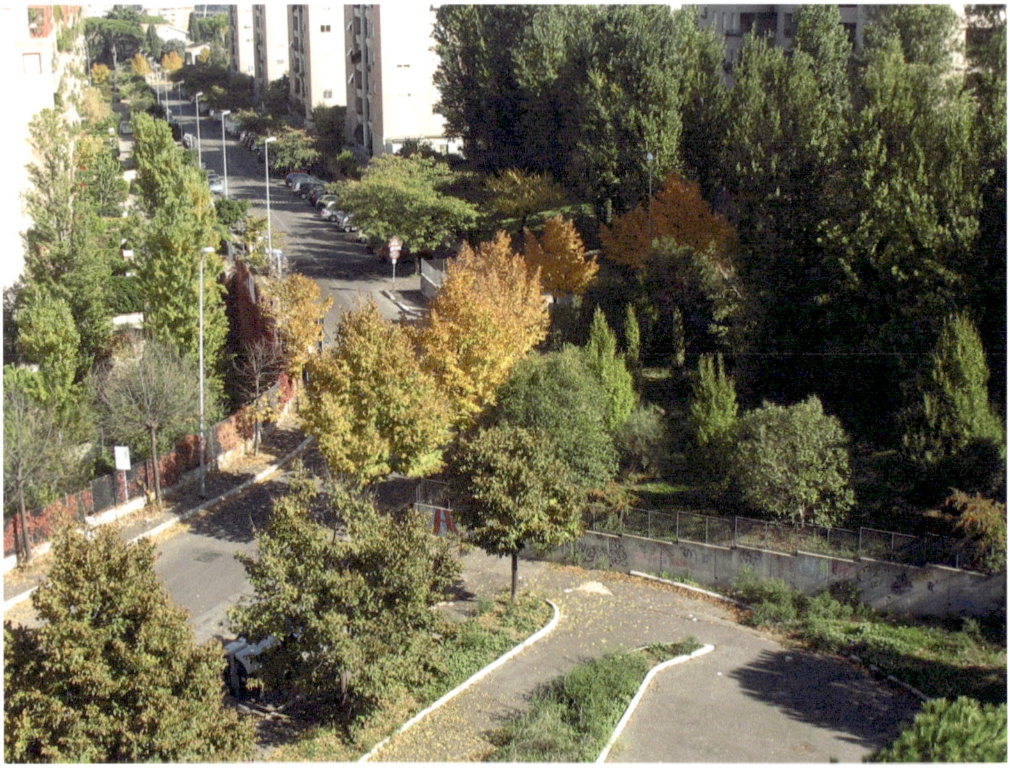

Figure 4. A photo by the Author of autumn in Via Dell'Orsa Maggiore, Rome, (November 2009).

“Speak to the earth, and it shall teach you”

Job 12:8

Chapter II

On Environmental Protection

Said a tree to a man: "My roots are in the deep
Red earth and I shall give you of my fruit"
And the man said to the tree: "How alike we are.
My roots are also deep in the red earth. And the
Red earth gives you power to bestow upon me of
Your fruit, and the red earth teaches me to receive from
You with thanksgiving."

The Wonderer
Kahlil Gibran
(1932)

II.1. A Birthday Present For a Nobel Laureate

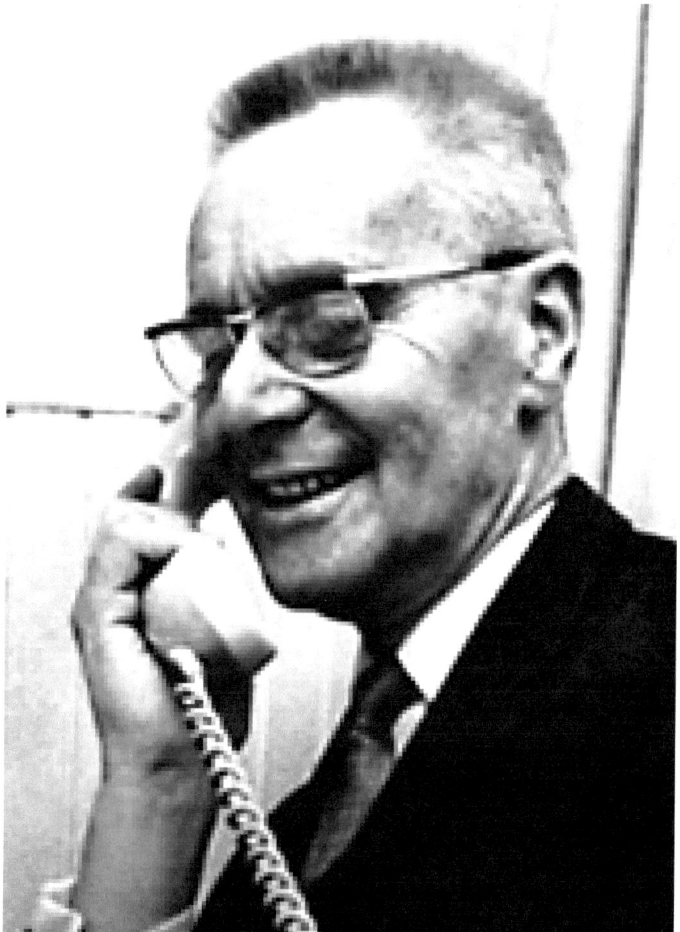

Professor Jan Tinbergen: The First Nobel Laureate in Economics; Rotterdam, (1971).

I was privileged to have had the opportunity to work closely with the late Professor Jan Tinbergen (1903-1994) at the Centre for Development Planning, at the Netherlands School of Economics (later was re-named Erasmus University) in Rotterdam, Holland. For over five years I assumed the position of a senior scholar and a staff member at the said School during the early Seventies of the last Century (1969-1974). During this period, I came to intimately know and highly appreciate not only the outstanding Professor Tinbergen, but also the most loving and kind human being. In this brief article, I shall share with you, my dear reader, some of my memories with this wonderful man.

Jan Tinbergen: The First Nobel Laureate in Economics

Professor Tinbergen started his academic career as a ***"theoretical physics scholar"***. Actually, in his early years he was a colleague of the famous ***Albert Einstein.*** Then, as he told us, he got tired of too many abstract mathematical and physics theories that he decided to shift to a practical subject. He chose economics.

Given his very solid and advanced mathematical background, and his very sharp and analytical mind, he was able to penetrate the subject of economics from a mathematical/logical perception. ***He was able to see in economics what other economists could not see.*** His deep insights into the subject was thus instrumental in coming up with his ***new*** theoretical yet pragmatic approaches for ***economic policy***. His outstanding contribution was extremely helpful in fixing the European economic crises after the World War II, and in helping many a developing country in their derive towards economic development and prosperity. His books and articles are full of mathematical reasoning, proofs and models. However, Professor Tinbergen has always told us, his colleagues and friends, that "***no matter how complex and sophisticated your mathematical modelling of an economic issue, concept or theory is, unless you are able to explain it in a simple and understandable language to the frequently non-economist policy maker, you have not succeeded as a policy advisor".*** And he had very sincerely and consistently adhered to this principle; a lesson that proved to be extremely helpful for me during all my years as a policy adviser at FAO and beyond.

The outstanding works of Professor Tinbergen included many books and numerous articles that can be followed, by the interested readers, in the literature on the subject. However, let me only highlight a few of his great achievements: first his ***dynamic models for economic development and policy analysis; and his planning in stages approach for economic policy***. Here, he has distinguished and very carefully elaborated three stages: ***the macro-stage; the sector-stage; and the micro-stage.*** The second is his ***semi-input output models.*** The two approaches are inter-related and enforce each other. Professor Tinbergen also emphasized such aspects as ***planning for the "poor", spatial development*** and ***decentralization approaches***. Professor Tinbergen's theories and policy prescriptions, as noted above, were instrumental in fixing the European Economies after World War II.

In recognition of his outstanding achievements, Professor Jan Tinbergen of Holland and Professor Ragnar Frisch (Econometrician) from Norway, were the ***first*** persons to be, jointly, granted the first ***Nobel Prize*** in ***Economics*** in 1969. They shared the prize ***"for having developed and applied dynamic models for the analysis of economic processes".*** Before 1969 there were ***no*** Nobel Prize awards for Economics as a discipline: Economics was not recognized as a "science". It was because of Professors Frisch's and Tinbergen's contributions to humanity in the field of economics that the subject "***Economics"*** was ***recognized*** as a ***"Science***" and a Nobel Prize was thus granted in this field only since 1969 until today.

Professor Tinbergen was not only an outstanding academic, but also a wonderful ***human being.*** He had a very powerful mind yet a very kind personality. His sharp mind developed deep perceptions of complex and intriguing issues and problems. His thinking was so clear, logical and systematic that renders unforeseen solutions to difficult problems easy to come up with. Meanwhile, my close encounter with Professor Tinbergen demonstrated to me beyond any doubt, that he enjoyed a very ***humble, pleasant, simple and loving character.***

A Birthday Present for Professor Tinbergen

Here is one of our encounters with Professor Tinbergen that demonstrates his very pleasant, non-orthodox and visionary character.

Once my colleagues and I noted that Professor Tinbergen's birthday was coming soon. So we decided to give him a present. Appreciating his very kind and loving character, we did not know what to give him. So we decided to ask him:

- "Professor, you know that your birthday is coming in a couple of weeks. And we, the staff members of the Centre, wish to give you a present. Please tell us what is it you would like us to bring you?

- He replied: "thank you so much; you should not have such a trouble".
- We replied: "there is no trouble at all; it is our honour and privilege to do so. Please tell us what is it that you would like us to bring you?"
- Being the always thoughtful Professor, he said; "Ok, let me think about it and I shall tell you tomorrow".
- The following day we went to him, asking the same question.
- He said: "My dear friends, I thought last night about your kind request. There is something I would very much like to do; if you can arrange it, I should be very grateful".
- "What is it Professor?" We all replied together. "We shall be delighted to do; please tell us."
- The Professor smiled and looked to us and said: "my friends, there is a "condition" you have to promise me that you will honour, before I tell you; do you promise?"
- "Yes, absolutely", we replied.
- He said: "my condition is that you do ***not*** approach the media, e.g., the TV, Press, photography, etc on my wish. You just arrange it quietly with the concerned authorities. Do you promise?" He added.
- "Yes of course we promise". By that time we were so exited: what is it that involves the authorities and should be denied to the Press and TV! "Please tell us Professor". We added.
- Professor Tinbergen smiled again, a smile that is as sweet as a child's, and said: "My dear friends, you know I do not own a car, I do not drive, and I use the tram every day for my transportation. Every day I observe the tram's driver with great admiration. I do like very much the "gold" like handles he manoeuvres to drive the tram. And my friends, if a permission is obtained from the Dutch Ministry of Transport for me to drive the tram (any tram) under the close eye of the driver next to me, I shall be very happy. And this will be the best birthday present I ever got. Do you think this is possible?"
- And we said "we shall try professor to fulfil you desire at any cost".

Our Dutch colleagues immediately contacted some people at the Ministry of Transport expressing the Professor's wish. Within less than a week, permission was granted for the professor to drive the tram for a distance between two stops under the close watch of the driver. The Ministry of Transport also honoured the Professor's wish for privacy: no press, TV or photography.

On the day of action, we joined Professor Tinbergen in driving the tram in Rotterdam for a distance between two stops. He was so happy and thrilled, like a child, during and after this exercise. Following his adventure, he provided his profound, sincere and extremely warm thanks to the driver of the tram. He thanked us so very much for succeeding in granting him what he called his "best" ever birthday present.

On Environmental Protection

Appreciating Professor Tinbergen deep thinking, a couple of weeks later, I asked him: What was his thesis behind his non-orthodox present? And he said: ***"the environment". This concern of Professor Tinbergen for the environment was in the early Seventies, when the subject was not on the agenda of the majority of neither the world's scientists nor the economists. Thus, by all accounts Professor Tinbergen was, indeed, a pioneer in that field as well as in economic policy and development planning.***

The Dutch Perfectionism

Reflecting on Professor Tinbergen's deep, thorough and perfectionist, yet simple and down to earth, way of life, his character is, indeed, in line with the Dutch perfectionism. To appreciate the Dutch perfectionism, look at the beautiful Dutch Tulips and the great art works of the outstanding Dutch Artists like Rembrandt and Van Gogh.

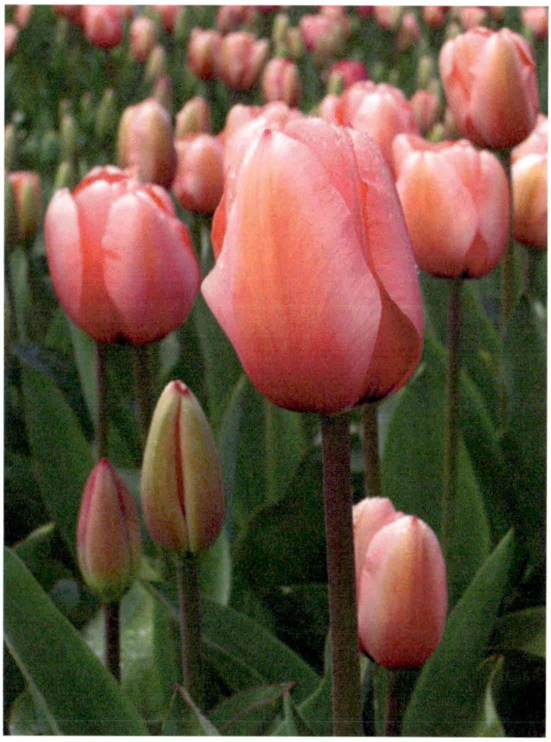

Dutch Tulips.

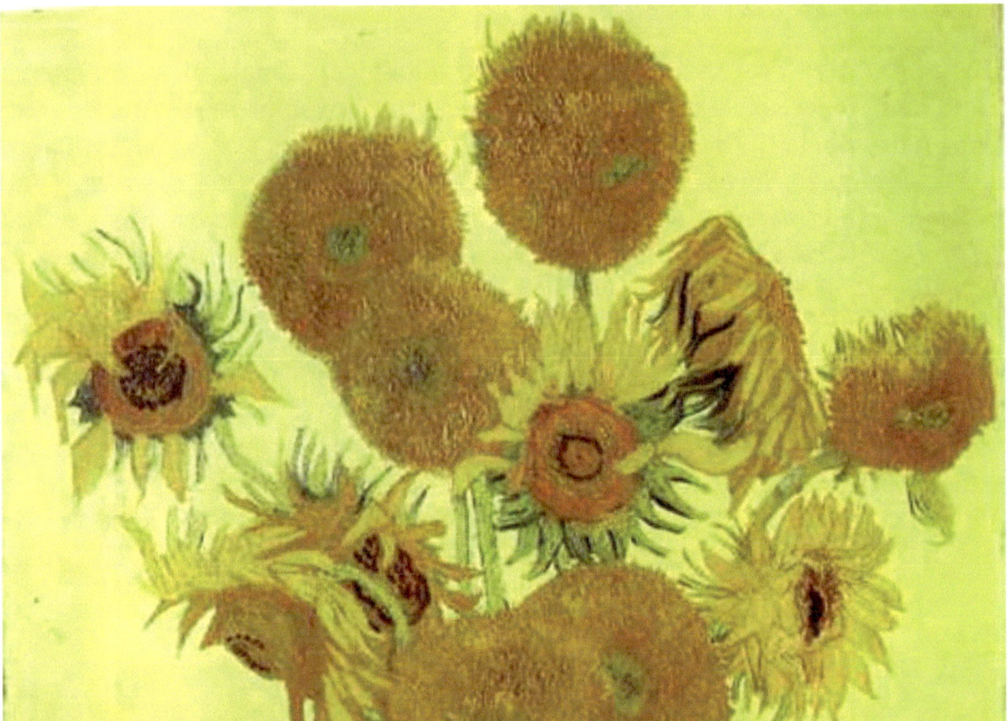

Flowers by the Dutch Artist: Vincent van Gogh.

II.2. Behold a Paradox:

Growth versus the Environment

This article was inspired by the recent (2010) environmental Gulf of Mexico's oil spill crisis in the USA.

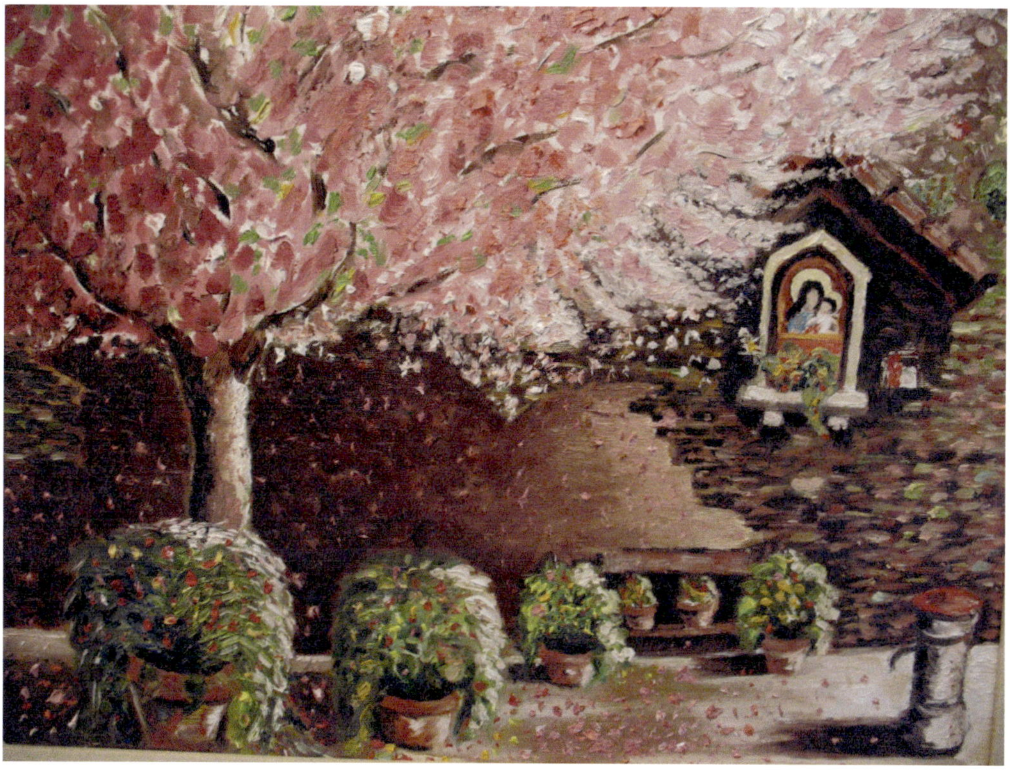

Figure 1. Environmental beauty: spring in Rome.
An oil painting by the Author (oil on canvas: 40X50 cm).

Introduction

Behold the following ***paradox***: it is the process of ***accelerated economic growth*** that often leads to ***pollution and natural resource degradation***; yet only through ***accelerated economic growth can environmental protection and natural resource conservation be sustainably maintained.***

This paradox can be illustrated by the on-going case of the environmental crisis of oil spills into the Gulf of Mexico in the US. Today (Mid July 2010), almost after three months and up to 184 million gallons of oil spills polluting the Gulf's water, only today some positive and promising signs are in the horizon for solving the problem; though the problem is not yet completely resolved.

Paradoxically, this environmental crisis was, indeed, the ***result*** of highly ***accelerated rates of economic growth*** in the USA leading to an insatiated appetite for energy coming from oil. Today, to the relief of millions of people along the Gulf Coast, the big, billowing brown cloud of crude, at the bottom of the sea, has disappeared (from the underwater video feed) for the first time since the disaster began in April 2010, as BP closed the last of three openings in the 75-ton cap lowered onto the well earlier this week.

The utilization of those most advanced technological innovations, (which are being tried for solving this huge problem) has only been possible through ***accelerated economic growth*** that led to the availability of the technology and the needed resources for its application.

Since accelerated economic growth is the culprit for the crisis, yet also for the solution, the ***question,*** therefore, should ***not be:*** whether to grow or not to grow? But rather ***how the process of accelerated economic growth can be so carefully designed and how relevant policy measures be rationally formulated in order that economic growth would conserve the natural resource base, protect the environment, be technically appropriate, economically viable and socially acceptable? Only then can the risk of such crisis be minimized.***

For embarking on the road towards answering those questions, the following is an abstract summary of some ***suggestions*** and ***lessons of experience*** with particular emphasis on lessons learned for ***developing countries***; and focusing on Sustainable Agriculture and Rural Development (SARD) in those countries.

Environmental Considerations and Natural Resource Conservation at the Macro-economic level

At this level, countries' national development strategies are elaborated. At this level, environmental considerations and natural resource conservation should assume high priority.

To this end, the following principles provide a starting point:

- First, national economic policy needs to be designed such that the ***present generations leaves future generations no poorer, and even better endowed, in the national assets than present generations.*** This is conducted while taking into account such factors as the likely population growth. Those assets include such components as: the natural resource base, knowledge and skills and the stock of man-made physical capital.
- Second, current patterns of population and economic growth need to be environmentally sustainable in the long run. Thus, conservation of the country's natural resource base assumes high priority among the macro-economic objectives.
- Third, for the conservation of the natural resource base, emphasis needs to be further attached to such aspects as: fostering institutional reform, enhancing people's participation, revisiting the role of environmental consideration in the education system at various levels, supporting the role of the private sector and revisiting the functions of the public sector, and other relevant socio-economic aspects of development.
- Fourth, in many countries national income accounts are deficient in the treatment of natural resources and the environment. Natural resources, unlike other tangible capital, are not treated as economic assets subject to depreciation with use. Consequently, losses of natural wealth are sometimes misrepresented as increases in income. Developing countries, therefore, need to revisit their national account systems to ensure that the countries' wealth of natural resources is reflected as economic assets subject to rational utilization and conservation.

Environmental Considerations and Natural Resource Conservation at the Sector level

The parameters and economic indicators resulting from the macro-economic analysis are made operational at the sector level of investigation. The important issue at the sector level is to examine the appropriate rates of growth for the economic sectors in the economy to achieve the national sustainable economic rates of growth ascertained at the macro-economic analysis. The analysis of the sectoral- interrelationships is usually carried out using the traditional Leontief input-output structure.

The incorporation of the environmental considerations into SARD taking into account the relationship of the agricultural sector with other sectors in the economy could be quantifiably analyzed using an extended input-output model.

The main approach for the extended input-output model (which has different variants by different authors) is that input-output models are based on accounting identities. However, the model could become more analytical through the introduction of certain assumptions about the nature of the production functions in the envisaged economic systems. In addition, material flows to and from the environment can be systematically included in the input-output model without disturbing the accounting identities. This is done through adding environmental activities which could basically include water, land and air. Furthermore, flows of matter from those activities to various sectors of the economy should be balanced by equivalent flows from different sectors to the environment.

The extended input-output model can next be solved to obtain such useful results for the agricultural policy maker:

i. ***Ecological/environmental Impact Tables***

Those tables show the effect on the ecological inputs and outputs of the sector of supplying one dollar's worth of economic commodity at producer's prices to final demand. Sectors and sub-sectors included in the input-output table can thus be ranked according to their ecological/environmental impact. Policy measures designed to minimize such impact could therefore be formulated on firmer and quantitative ground.

ii. ***Scenarios for Investment in Anti-pollution and Anti-depletion of Natural Resources:***

Given the society's "tolerance" level to pollution and environmental degradation, the extended input-output model can also estimate the required level of investment to achieve certain environmental improvement. Sensitivity analysis could be further used to develop scenarios for investment alternatives corresponding to different tolerance levels. The resulting quantitative function helps the concerned policy maker in examining possible options and tradeoffs.

Environmental Considerations and Natural Resource Conservation at the Project level

The traditional cost-benefit approach for project analysis is mainly based on the estimation of the internal rates of return on invested capital. If the internal rate of return exceeds a certain level, the project is deemed viable and acceptable. This approach, therefore, implies that the analysis tries to maximize the productivity of ***capital*** as the limiting factor.

However, for sustainable development and for environmental considerations, the ***natural resource base*** becomes the limiting factor for production and not only capital. Thus, approaches for project analysis need to be revised to favor those projects that maximize the productivity of the natural resource base used in the project.

To this end, FAO suggested that ***"With some modification, environmental impacts can be incorporated into cost-benefit analysis. Three main adaptations can be used to accomplish this goal. First, socio-economic cost-benefit analysis often uses shadow prices to estimate the long-term effects on the environment. This should be done regularly, even if it is (difficult) impossible to make accurate estimates of environmental costs and benefits. As a by-product of their main functions, environmental impact assessments provide useful data for making such estimates. Second, cost-benefit analysis may include safe minimum standard criteria such as those widely used in engineering designs. Safe minimum standards could specify the environmental criteria that development projects should meet, although natural resource system are usually highly complex and appropriate criteria are difficult to specify. Any additional cost required to satisfy the standards could be added to other project costs in the analysis. Third, while it would be unrealistic to expect every development project to have a positive environmental impact, this could be insisted on at the programme level." (See FAO, State of Food and Agriculture (SOFA), 1989 p 81.)***

Some Concluding Remarks

SARD is a process not an event, and environmental considerations and natural resource conservation should be integral components of that process at all its evolving design and implementation phases. While this approach can minimize the opportunity of environmental disasters, it will not eliminate such unfortunate events. When that happens, its human and financial costs will be enormous, yet with further accelerated economic growth, assistance of the international community and patience overcoming such disasters will be feasible.

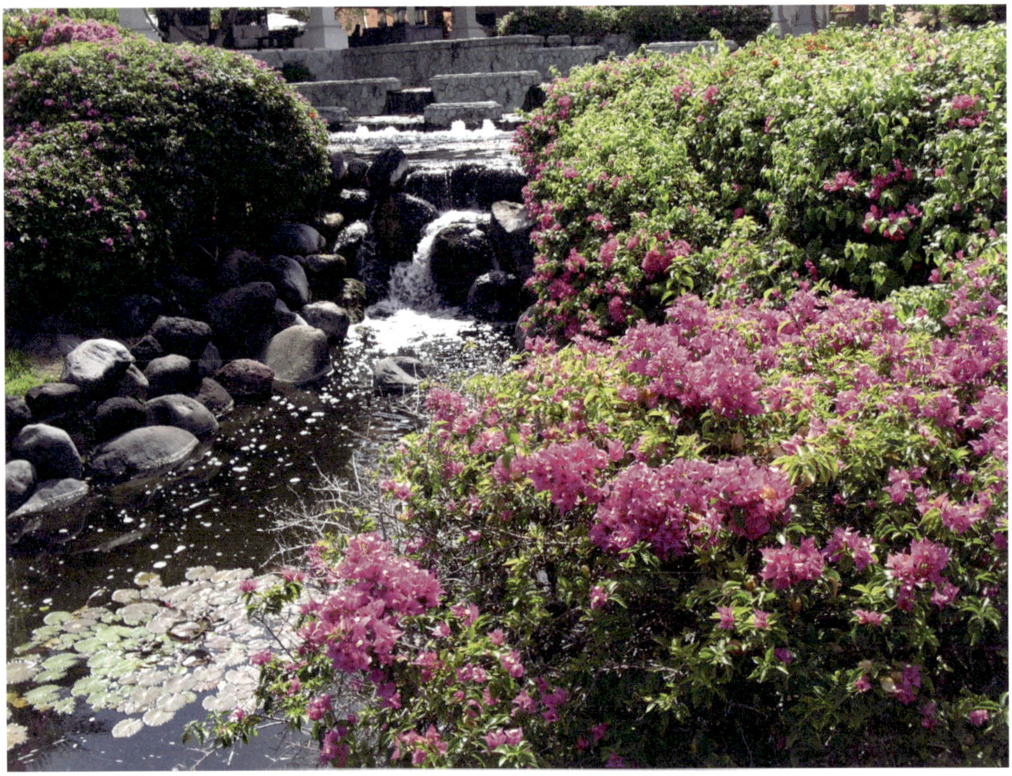

Figure 2. Beauty of the natural resource base: A photograph by the Author in Bali, Indonesia

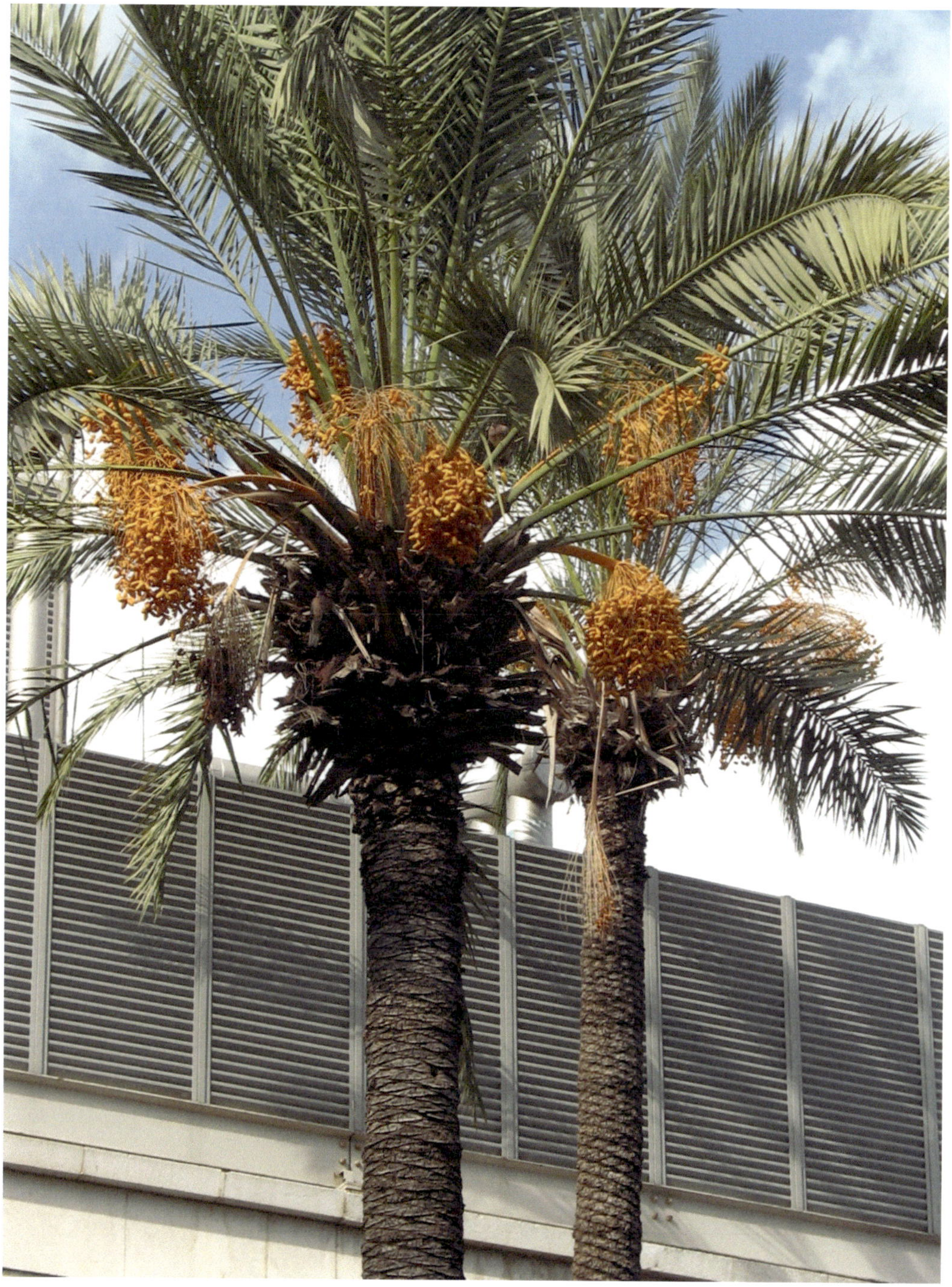

Figure 3. Urbanization versus the palm tree. Photograph by the Author in Beirut, Lebanon.

II.3. Towards Sustainable Agriculture and Rural Development in China

Lessons of Experience

'The miraculous success of the Chinese economy during the last two decades is remarkable. One of the country's great achievements in agriculture has been in working towards Sustainable Agriculture and Rural Development (SARD) through Chinese Ecological Agriculture (CEA). Many a developing country is facing this difficult challenge. Based on his FAO's working experience in China, the author summarizes a number of lessons learned from China in achieving this goal. Those lessons, however, must be tailored to fit the existing conditions, challenges and aspirations of concerned countries prior to implementation."

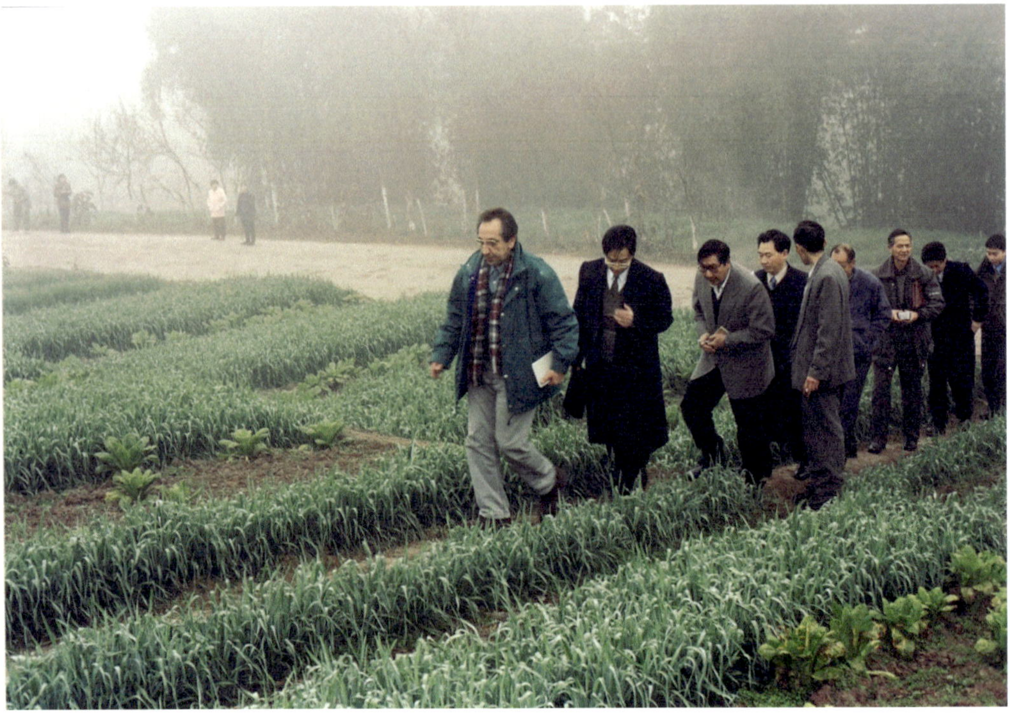

Figure 1. The auther with FAO and Chinese colleagues at an FAO project's site in Sichuan China.

China is currently the world's second largest economy after the United States. The country's purchasing power parity was $8.77 trillion in 2009. China is the world's fastest-growing economy, with an average ***growth rate of about 10% for the past 30 years***. In 2009, China's per capita income was $6,567 (IMF). China is also the second largest trading nation in the world and the largest exporter and second largest importer of goods.

China is one of the world's largest producers and consumers of ***agricultural products.*** Virtually all arable land is used for food crops. China is the world's largest producer of ***rice and is among the principal sources of wheat, corn (maize), tobacco, soybeans, peanuts (groundnuts), cotton, potatoes, sorghum, peanuts, tea, millet, barley, oilseed, pork, and fish***. Major ***non-food crops***, including ***cotton, other fibers, and oilseeds***, are also produced in China but constitute a smaller proportion of the country's foreign trade revenue.

Yields in China are generally very high due to intensive cultivation. China ranks first worldwide in farm output per hectare in many agricultural products. In 2003, China ***fed 20 percent*** of the ***world's population*** with only ***7 percent of the world's arable land.***

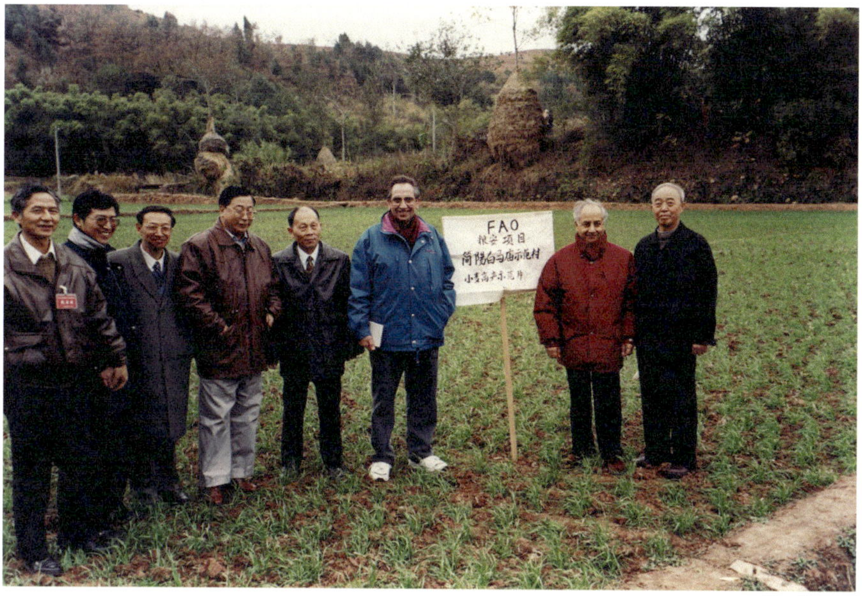

Figure 2. The auther with Chinese colleagues at an FAO project's site in Schuan China.

SARD and CEA in China:

The ***Chinese*** drive towards SARD through CEA is based on a ***comprehensive approach for the incorporation of environmental considerations into agricultural development***. That is, the main characteristics of CEA and SARD include an agricultural production system that adapts itself to enhancing the ***environment, conserving the natural resource base,*** maintaining the balance in the ***ecosystem;*** entailing an ***optimum agricultural system*** with ***sound social, economic, ecological and environmental effects.*** The Chinese model for SARD and CEA ***integrates*** sound practices of ***traditional*** Chinese agricultural ***technology*** with ***modern science and technology*** to create ***sustainable agriculture and rural development*** in the country.

There are many carefully adapted ***indigenous and modern techniques*** which are commonly utilized in ***CEA farming systems*** in China to achieve ***sustainable agriculture and rural development (SARD).*** The following are some key examples:

- Using of ***labor- and technology-intensive cropping systems*** (often emphasizing complex inter-cropping systems with high-value crops, using large amounts of organic manure);
- Developing ***rural energy resources*** including ***biogas, wind energy as well as other energy saving techniques but used on a small-scale by households*** (this is designed to utilize seasonally abundant and surplus labor);

- Adopting effective ***dry-land framing technologies***, including such techniques as deep plowing, soil moisture conservation, the use of straw and stock manure for soil and moisture conservation, etc.;
- Introducing ***integrated pest management*** methodologies, including techniques which combine the use of limited amounts of farm chemicals applied at specific times along with utilisation of natural methods, such as crop rotations and the use of other natural, labour-intensive methods;
- Utilizing ***integrated plant nutrition management approaches***, including a combination of applying chemical fertilisers according to technically efficient timing schemes and combining them with applications of organic manures and recycled plant wastes;
- Incorporating ***anti-soil erosion technologies*** which use integrated systems of combining land terracing, the conversion of land from cultivated land to pastures and forests, the extensions of sustainable forest and livestock management systems;
- Developing, treating and utilizing ***sewer water irrigation schemes***, using urban and industrial waste water to safely irrigate land in areas that are extremely water short. In addition, this includes the use of treated waste water for increasing the organic nutrients in the soil; use of physical recycling; the integration of multi-cropping and pond fisheries systems into single crop rotation systems where the latter exits;
- Raising of fish and ducks in paddy fields.

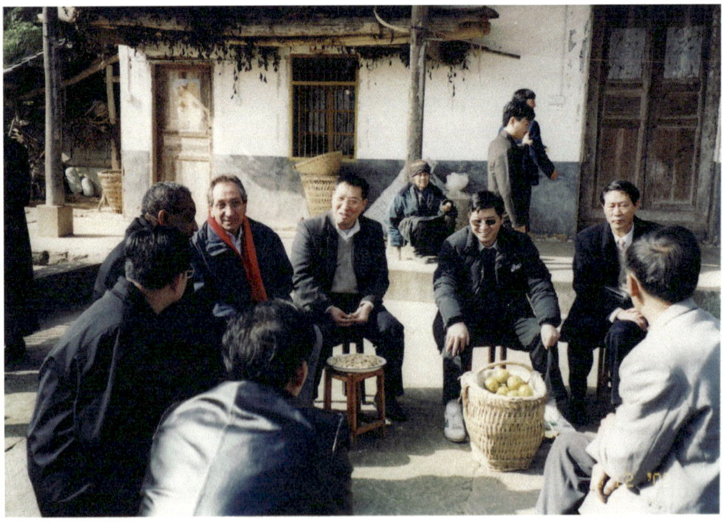

Figure 3. A Chinese farmer (third from left) discussing his actual technological and socio-economic constraints with concerned local officials and FAO team: a highly participatory approach.

Lessons of Experience

With the above in mind, the following are selected lessons learned from the Chinese model:

Resilient and Endogenous agricultural technology

- The first factor for the success of China's CEA and SARD programmes is mainly inherited in the resilient and dynamic attributes of traditional Chinese agricultural technologies ***(indigenous) based on a highly participatory approach. Endogenous SARD models that are based on appropriate technologies that fit within the concerned country's traditions and cultural background are, thus, having the highest chances of success.***

Integration of traditional with modern technologies

- Secondly, traditional farming systems in China have always ***combined conventional techniques with modern*** (sometimes transferred from within the sector and/or from abroad) ***agricultural technologies***. Chinese producers have always created technological systems on the basis of each area's unique set of factor endowments. Those carefully designed enterprises produced an agricultural system that is highly productive yet causing relatively low stress on the environment. In general, the techniques in the CEA's farming system encourages ***low material input, recycling, high labour use, and balancing the level of modern inputs with traditional agronomic practices. Therefore, careful integration of traditional and modern agricultural production techniques is another factor for efficient and pragmatic SARD.***

Policies for sustainable agriculture and rural development

- The third key factor for the success of CEA and SARD approaches in China is the careful design and implementation of a comprehensive and complementary set of ***policies for sustainable agriculture and rural development,*** and the incorporation of those policies along with appropriate technology into the national planning process. The incorporation of SARD dimensions into agricultural policy and planning in China has been focusing on the following measures:

  - Integration of ***growth (efficiency), equity and sustainability*** within a coherent and operational framework;
  - ***Incorporation of SARD objectives into the national planning and objectives*** - in particular: contribution to GDP; population and employment; improved income distribution and poverty alleviation; balance of pay-

ments; and environmental protection and natural resource conservation; and

- ***Integration of SARD dimensions into the agricultural development process;*** in particular, in the national development strategy developed at the macro-stage; in the multi-sectoral analysis developed at the sector-stage; and in the extended scope of project analysis to incorporate sustainable development and environmental protection criteria developed at the project-stage.

Minimization of the depletion and degradation of the natural resources base

The critical elements incorporated in the formulation of ***policies*** and technological innovations for sustainable agricultural and rural development that simultaneously minimise the depletion and degradation of the natural resources base in China included, *inter alia*:

- Enhancing the development and access to ***modern (efficient) agricultural technologies;***
- Strengthening efficient ***marketing systems*** that provide producers with the incentives to engage in sustainable agricultural production and marketing activities;
- Stressing technical and economic support to areas known to have high ***potential*** for sustainable agricultural production;
- Enhancing local ***farmers' participation*** in CEA and SARD programmes through such measures as the "contracting systems" and market incentives;
- Creating and disseminating new technologies through strengthening applied ***research*** and efficient ***extension*** services;
- Expanding the scope of agricultural system to stress ***integrative practices***;
- Stressing ***regional decentralization***;
- Developing and implementing measures for ***poverty alleviation***;
- Formulating and implementing complementarily ***rules and regulations***; and Setting up an efficient national and regional institutional framework for the formulation and implementation of CEA and SARD programmes.

A highly participatory approach

Another essential factor for the success of the Chinese model has been the farmers' highly participatory approach. On a continues basis, farmers meet among themselves and with concerned Governmental, non-governmental (NGOs), FAO and other concerned national, regional and international bodies, as appropriate, to discuss their challenges and come up with operational solutions. To stimulate this fruitful process, certain non-monetary incentives are provided by GOC to local farmers. For instance, Figure 3. Below shows a local Government and FAO joint meeting in Sichuan where selected four ***local farmers*** who have ***achieved substantial increases*** in their farm productivities were ***honored*** and received honorary certificates.

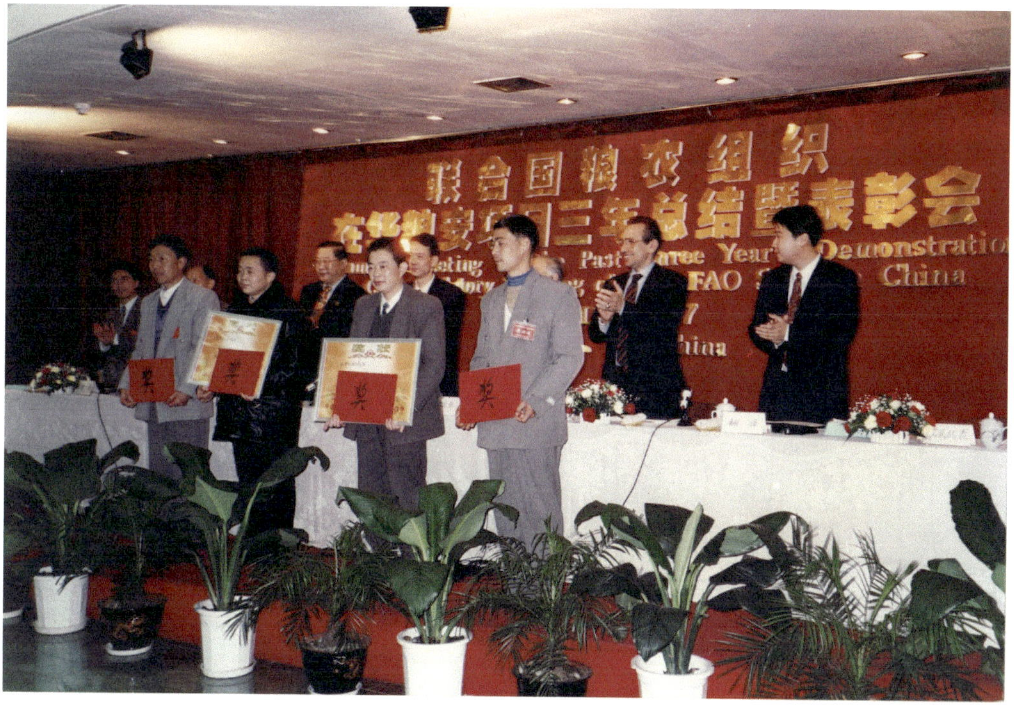

Figure 4. Honouring four outstanding local farmers in Sichuan.

“We love things that bear the mark of grime, soot, and weather; and we love colors and sheen that call to mind the past that made them.”

Tanizaki Junichiro.

Chapter III

On World Landmarks

"Once there lived a man among the hills who possessed a statue wrought by an ancient master. It lay at his door face downward and he was not mindful of it. One day there passed by his house a man from the city, a man of knowledge, and seeing the statue he inquired of the owner if he would sell it. The owner laughed and said: "And pray who would want to buy that dull and dirty stone?" The man from the city said: "I will give you this piece of silver for it." And the other man was astonished and delighted. The statue was removed to the city, upon the back of an elephant. And after many moons the man from the hills visited the city, and as he walked the streets he saw a crowed before a shop, and a man with a loud voice was crying: "Come ye in and behold the most beautiful, the most wonderful statue in the world. Only two silver pieces to look upon this most marvellous work of a master."Thereupon the man from the hills paid two silver pieces and entered the shop to see the statue that he himself had sold for one piece of silver."

The Wonderer
Kahlil Gibran
(1932)

III.1. When the Great Pyramid Spoke

"Had I not been solid, the Basilica of Saint Peter in Rome, the Cathedral of Saint Paul in London, the Westminster Abbey, and the Cathedrals of Florence and Milan all together could have found room inside me"

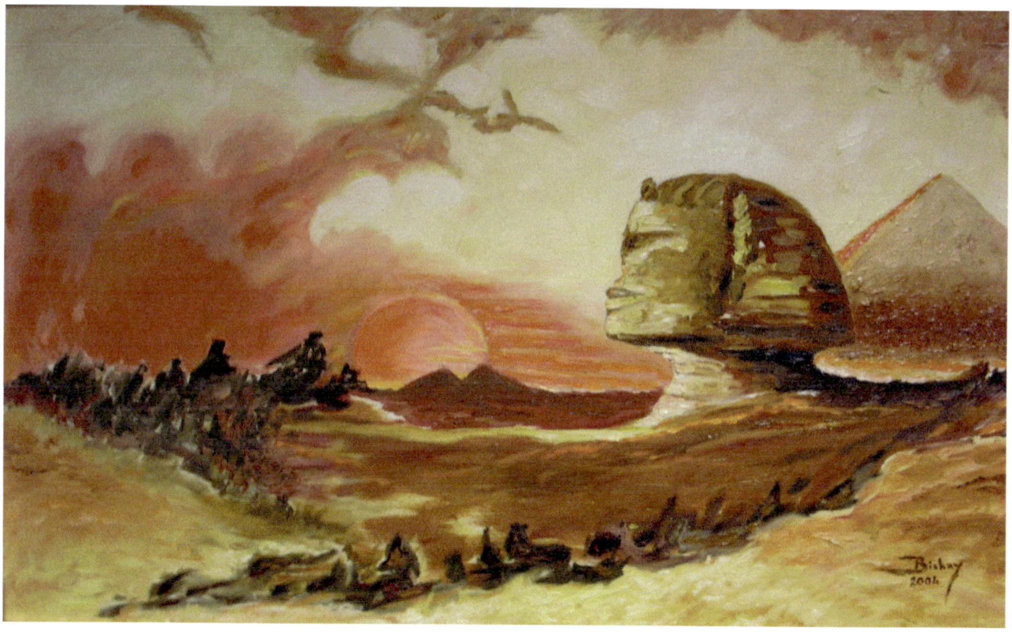

Figure. 1. The Great Pyramid of Giza guarded by the Sphinx. A portrait by the author: (oil on canvas: 70x110 cm)

The Great Pyramid spoke only once. He said: "I am The Great Pyramid of Giza, Egypt. I was born in the year 2629 BC (the start of work on my construction was in the year 2589 BC and it has taken about 40 years to complete). I am, therefore, about ***4639 years old***. I was built by the ingenious ***Pharaoh of Egypt: Cheops*** (that is his Greek name; in Arabic Khufo). I am so gigantic that all the above mentioned outstanding Basilicas and Cathedrals could have all been accommodated within my enormous structure had I not been solid. But thanks God I am solid, otherwise I could not have lived so long. Yes, I am so huge, indeed: My original height was 147 m; today I am ***139 m high***; that is the height of a building of about 46 stories. My base is a square. The length of ***each side is 230 m.*** Thus the area of my ***base is 52900 square meters***. My base perimeter is 921 m. My Ancient Egyptian builders used about ***2.3 million blocks*** of stones, weighing on average ***2.5 tons a piece*** for my construction. Thus my weight is about ***5.75 million tons***, or the weight of over ***6 million cars***. I am grateful for God's blessing on Egypt that I was built on such a solid hill in Giza which could withstand my weight for so many years, without resulting in natural disasters such as earthquakes. But as I have been standing here for thousands of years before Christ, I have witnessed the ***Holy Family's visit*** to my country: Egypt; and have observed them bestowing their blessings on my beloved land. Thus, I am confident that I shall safely live here for good", (Figures 1).

The Pyramid next revealed how He ***looked thousands of years ago*** from the outside, and what is inside him today. And He said: "When I was built, there was a ***complete outer casting*** limestone which covered all of me. There are still a few in places near the base. There were also carved ***inscriptions*** on those smooth surfaces of mine. It is estimated that those inscriptions could fill more than ***ten thousand*** (10 000) ***pages***. It is a pity that those inscriptions vanished when my external limestone casting was removed in the 13th century. And the stones were used in building homes. Since then I look as I do today. My current visible stones are arranged in rows. When I was built there were ***210 rows of stones***. Over the centuries, the upper seven rows were lost, leaving ***203 rows of stones today,*** (Figure 2). As regards what is inside me, you can enter me from the current opening at 17 meters above the ground. From there you will start a descending passage for 105 meters long and leading to an ***underground chamber***. However, if you climb up a narrow steep ascending passage for 39 meters you will reach another passage leading to my ***King's chamber***. This is my focal point: ***my heart***. That is where the King has been resting. The design and construction of the King's chamber is another engineering miracle. Above the King's chamber (dimensions: 10.8 x 5.7 x 5.2 meters) there are 7 rows of ***granite*** stones, each row is 7 cm in from the row below it. There are two ***airshafts*** (air passages) that connect the chamber to the sky outside on my north and my south sides. The Northern shaft gave way to the ***soul*** of my ***King*** to ***ascent to Heaven***. For that noble objective, my builders worked so hard on the design and construction of those shafts", (Figure 5).

The Pyramid was silent for a thousand years, and then He spoke again. And He said: "My sons and daughters of the modern world, here is a ***riddle*** for you: ***How, so long ago, was I perceived, designed and constructed to become perhaps the largest man-made structure in the world today?*** What architectural technology was used in my construction that was available for more than four thousand years ago, that was so simple (to be available) yet so powerful to accomplish such an achievement?" He then posed and smiled. And when we, the sons and daughters of Egypt, pressed him for an answer, He said: "This is my ***secret***. I can not reveal it now. My ***sons and daughters*** of the ***world*** follow the footsteps of the Pharaohs, and ***think, think and think*** again to resolve this riddle".

And since then scholars from all over the world worked along with their Egyptian counterparts on trying to resolve this riddle. They visited the Pyramid, carried out very accurate measures using the state of the art in technology, and applied the most advanced scientific approaches in architecture, physics and astronomy in trying to resolve the riddle: How the Pyramid was built? Thus, many hypotheses and theories were postulated by prominent Egyptian as well as world architects, physicists and astronomers to answer this question. But what astronomy has got to do with the question? Yes, astronomy proved to be relevant for the in-depth research on the riddle revealed that ***astronomy*** was also utilized in the ***orientation and measurement*** of the design and building of the Pyramid.

Of the many theories on how the Pyramid was built, a plausible one was provided by the French architect ***Jean-Pierre Houdin***. Houdin divided the building of the Pyramid into two phases: The first phase was the construction of the pyramid itself. The second phase was the building of the King's chamber whose base lies at a height of between 43 and 67 m from the ground. The challenge of the second phase was: how to lift from the ground 63 ton granite beams for the ceiling of the King's chamber?

Houdin's theory is that both phases were realized through two ***ramps:*** an ***external ramp and an internal ramp.*** The external ramp reached a height of 43 m and slope of 7%. The external ramp was constructed from the beginning with its full length. At that height, 73% of the Pyramid's volume was achieved. The external ramp was also used to bring up the heavy granite slabs (63 tons each) that covered the ceiling of the King's chamber. After the King's chamber was built, the external ramp's role was accomplished, and it was dismantled and its stones were used in building the rest of the Pyramid. The rest of the Pyramid (above 43m) was build using an ***internal spiral ramp***, as shown in Figures 3 and 4.

The pyramid next ***reflected*** on His ***eternal life***, and then He said in a voice like ***thunder*** for ***He was angry***: "It is ***not*** true that, as a tomb, I was built to glorify death. I am created to ***celebrate life*** after life. The Pharaohs deeply believed in the second life, and I was created as a ***custodian of the new life***. ***Nor*** is it true that I was built by slavery in Ancient Egypt. For my design and construction were so majestic and my realization was indeed a miracle; and miracles can only be performed by love and faith. Yes, my construction was ingenious and called for very hard work. But, that was done by the Egyptians with love and as an act of worship to their Pharaoh Cheops: Who was not only their King but also their God. Actually, there was a socio-economic policy objective in my construction: ***Work*** on me was performed during the summer ***flooding seasons*** of the ***Nile*** when farmers were idle, and thus my construction generated ***employment*** opportunities for them during those slack periods".

And ***He concluded***: "***I was built by an act of faith, love and worship, and so I shall be perceived by generation after generation.***" And He was silent again. I heard the Pyramid and was humbled by His Majesty and His human face.

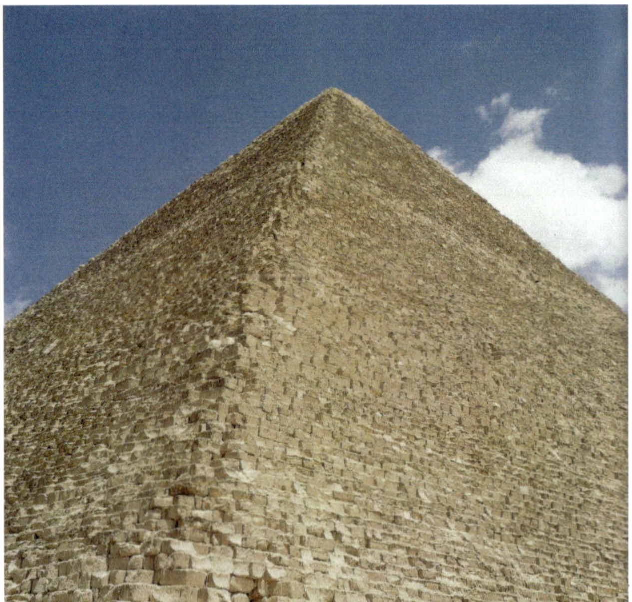

Figure. 2. The mighty Great Pyramid of Giza.

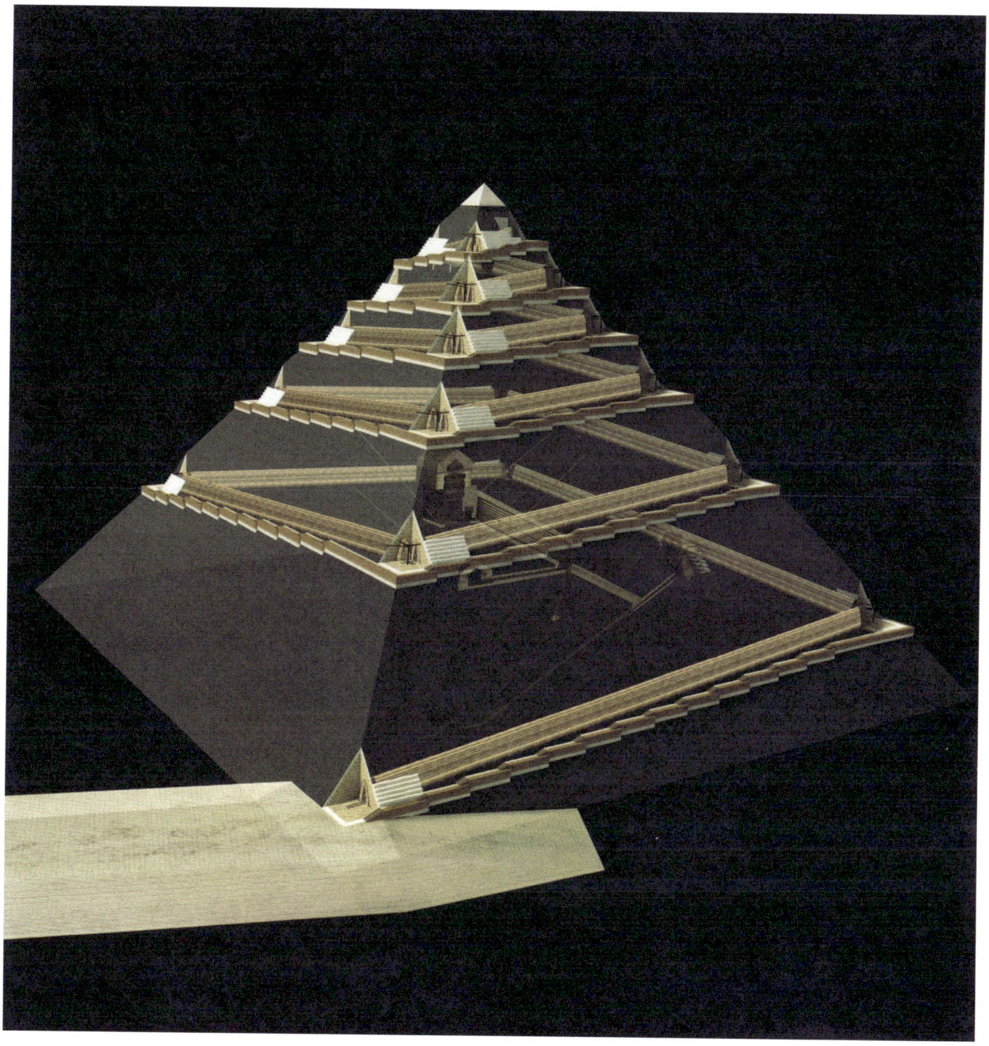

Figure. 3. Jean-Piere Houdin's theory for building the Pyramid: The external ramp

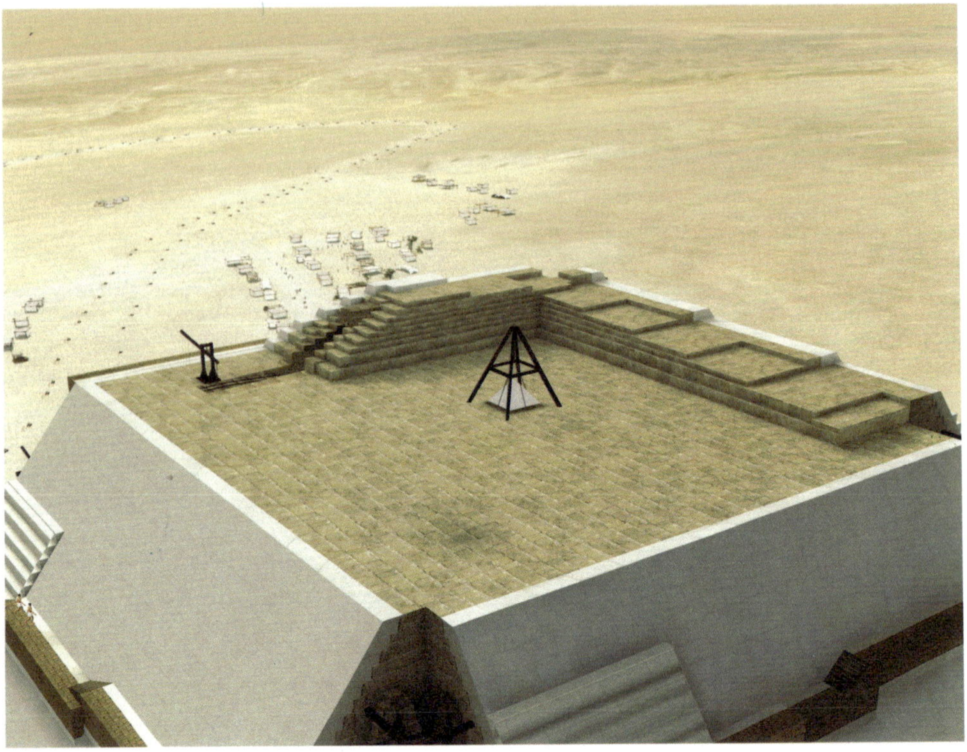

Figure. 4. Jean-Piere Houdin's theory for building the Pyramid: The internal ramp

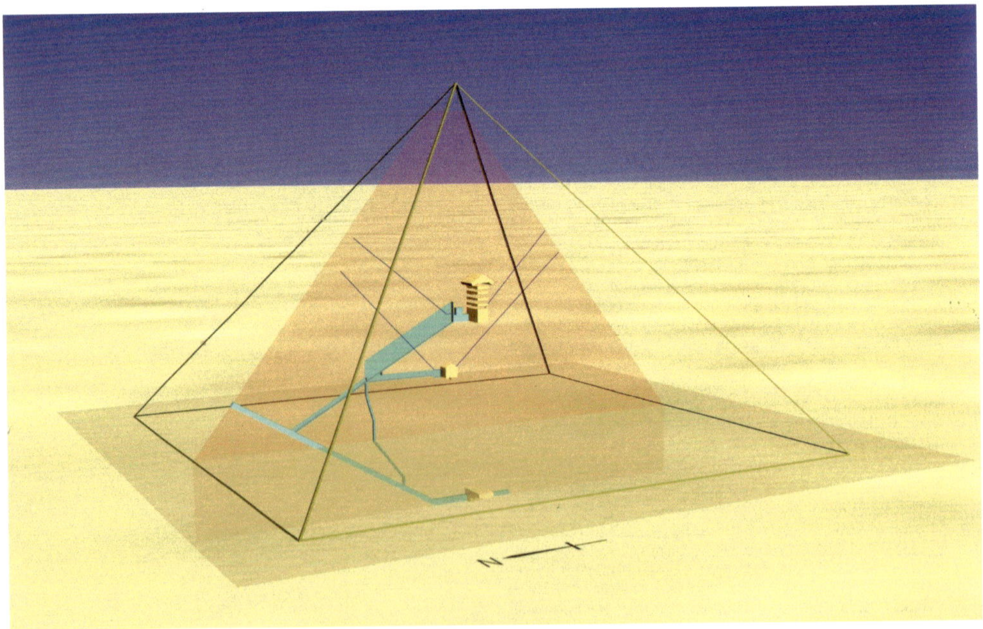

Figure. 5. A computer image of the inside of the Pyramid.

III.2. When the Nile spoke

"I am a river of life running from ex-eternity to eternity"

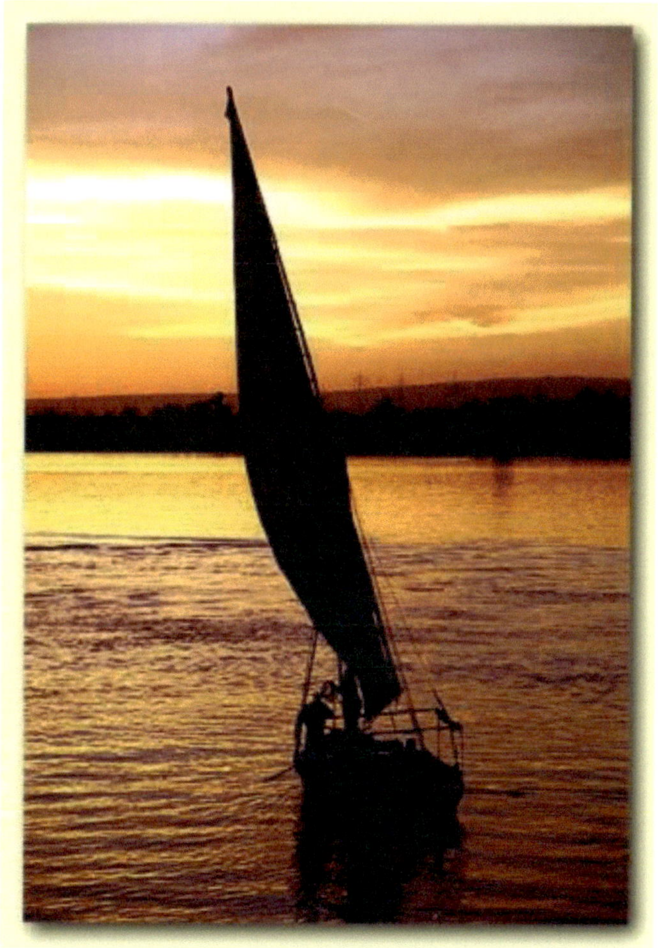

The Nile at sunrise.

The Nile spoke only once. He said: "I am the Nile River. My name was given to me by the old Egyptians: the Pharaohs. My name came from the Greek word "*Neilos*" meaning valley. As I used to flood the land in Egypt (prior to building the Aswan High Dam) leaving behind black sediment, the ancient Egyptians also named me the river "*Ar*" meaning the black river, a name that got forgotten over the centuries.

I am the longest river in the world: I run for *6695 kilometers* (4184 miles). I flow through 9 countries: Ethiopia, Zaire, Kenya, Uganda, Tanzania, Rwanda, Burundi, Sudan and Egypt. I do not believe in boundaries between countries, nor dividers between farms. I give my priceless *water,* the source of life, to all the people in those countries without any distinction. Like the sun, that gives its light and warms to all, I do. I do not distinguish between the good, the bad or the ugly. It is in such generosity in giving that God's teaching of love and humanity is practiced. Thus, it hurts me very much when I see a man fighting his brother for material disputes.

I do give with love my water for drinking, irrigation, transport, electricity and recreation. The wealth of fish in my water I give without reservation to the needy and the wealthy alike. True, I have some Nile Crocodiles swimming in my water, especially in the Sudan. But even those creatures are mighty and beautiful in their own way. Those animals have settled along my fertile banks for thousands of years. Most of my Nile Crocodiles are approximately 4 meters in length, although some have been reported as longer. The animals make their nests along my banks where the female may lay up to 60 eggs at one time. Some three months later the babies are born and are taken to my water by their mother. They remain with her for at least two years before reaching maturity and enjoy swimming in my water. Yes, those crocodiles kill to eat, but do not all humans do? However, when man kills his brother for the sake of political or power ambitions that is when I weep; and my tears run for thousands of miles and thousands of years.

Yet, I am grateful for I have been honored by many of the inhabitants on my banks. For instance, the ancient Egyptian calendar was made up of 12 months of 30 days each and was divided into three seasons based upon my cycles. Herodotus, the great Greek philosopher, wrote of me: "the Nile River rises of itself, waters the fields, and then sinks back again; thereupon each man sows his field and waits for the harvest." The great historian also called Egypt "the gift of the Nile". Today, exotic and sophisticated cities like Cairo grace my banks".

The Mighty Nile then waited for a thousand years and then He concluded: ***"I am a river of life running from ex-eternity to eternity; a drop of my water is the ocean and the ocean is a drop of my water".*** I heard The Nile but did not understand what He said. But He became silent again.

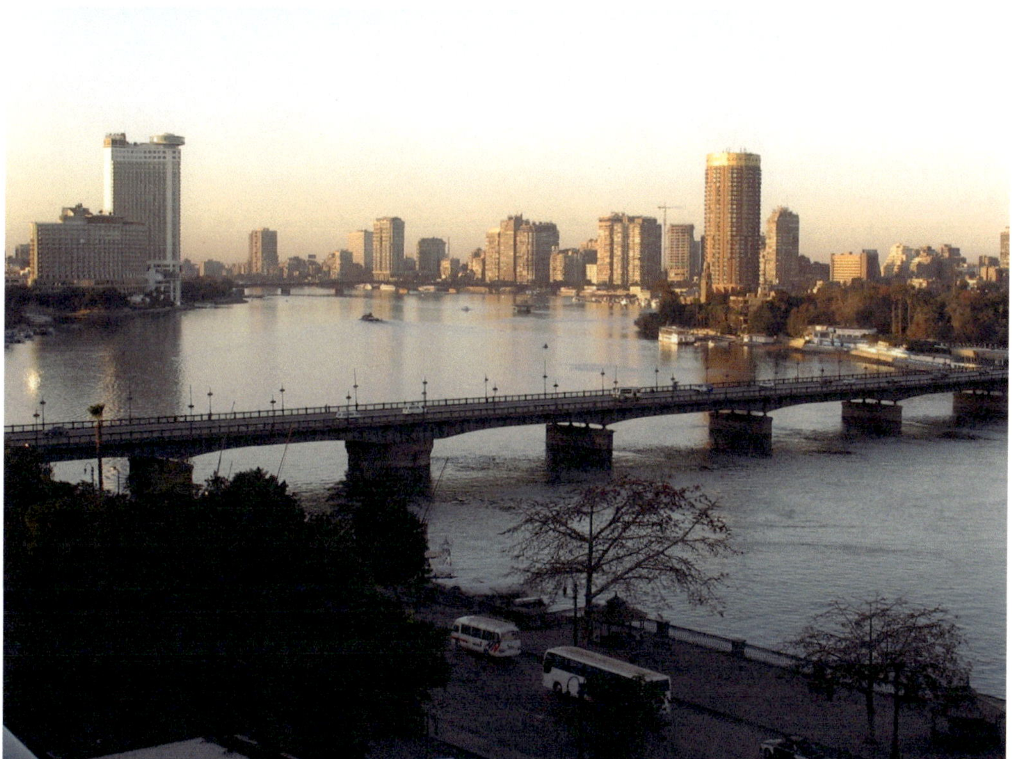

The Nile in Cairo (a photo by the author).

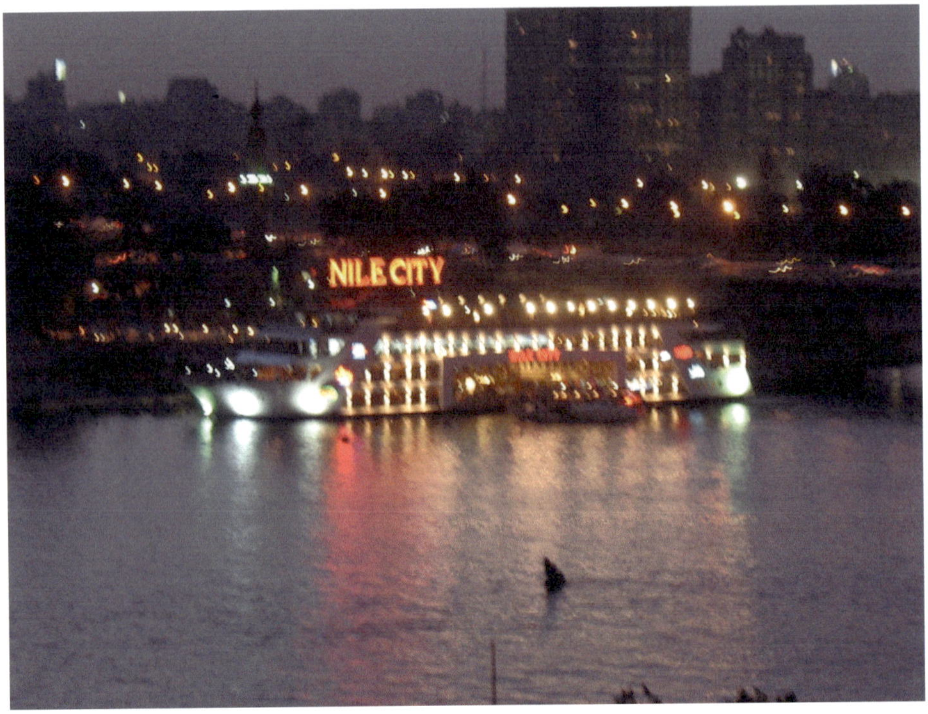

The Nile at night (a photo by the author).

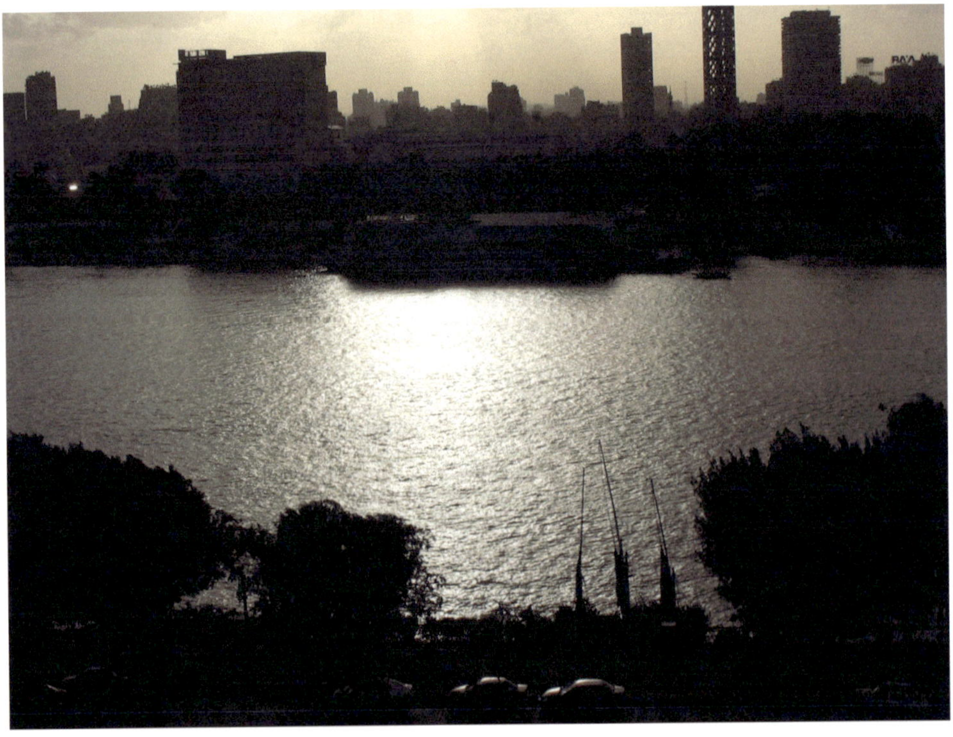

The Sun reflection on the Nile (a photo by the author).

III.3. Coptic Icons from Egypt[2]

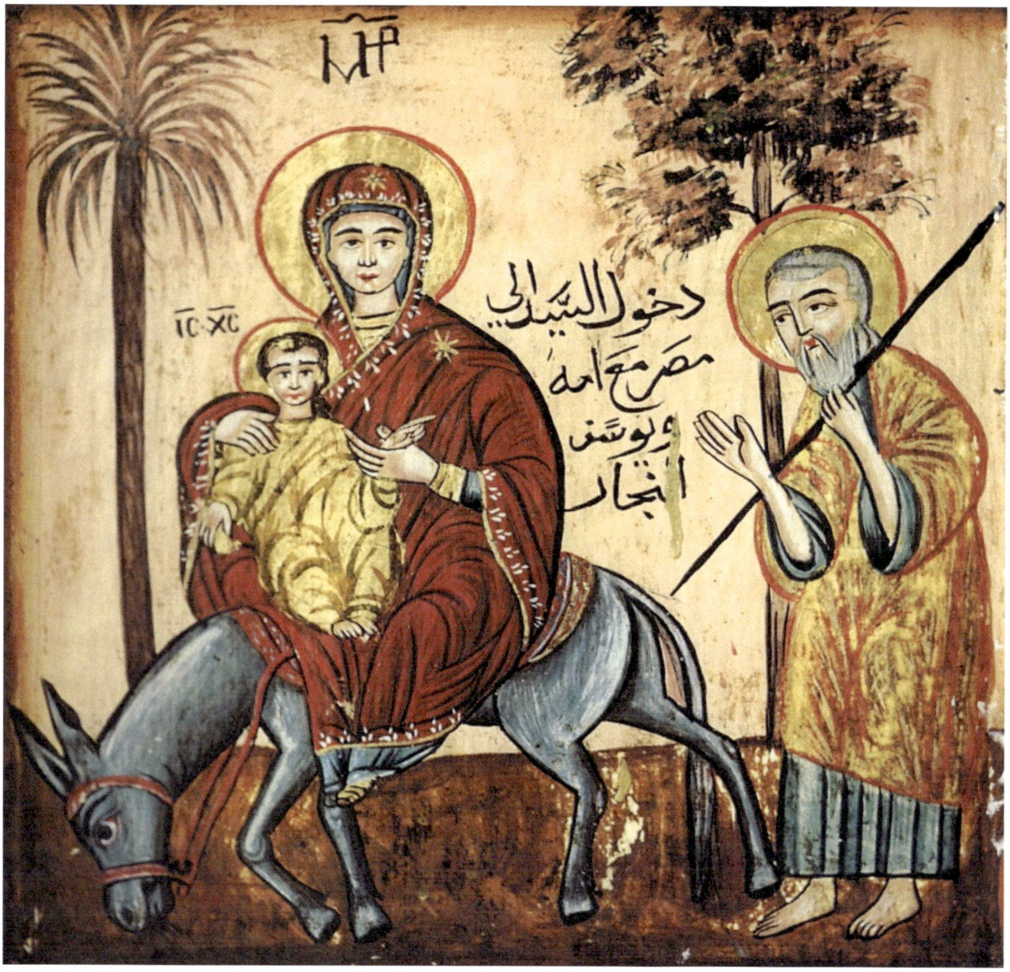

Icon 1. Entry into Egypt: The Holy Family's Journey to Egypt. (Note the simplicity of the work, yet the effective expressions in the eyes, even those of the donkey.)

---

[2] Selected sources:

- Orbis Terrae Aegypyiae, "Coptic Icons: Parts I and II", Lehnert & Landrock, 1998, Cairo, Egypt.
- M. Simalka, "Guide to the Coptic Museum" (in Arabic), 1930;
- Victor Awadallah, "The Icons of the Coptic Museum" (in Arabic), 1965, Cairo, Egypt.

The word ***"Coptic"*** is derived from the Greek word *"Aigyptios"* meaning the original inhabitants of a particular country, in this case Egypt. Thus the word Coptic actually means Egyptian. However, over the past centuries the word Copt in Egypt has been used to mean Egyptian Christian; and more specifically Christian Orthodox. Thus, the Coptic Church is generally understood to be the Egyptian Christian Orthodox Church.

The word ***"Icon"*** is also of Greek origin, and initially it meant a *"portrait"*. But, again over the years, it changed its meaning into Christian paintings. Coptic icons refer to the old Egyptian Christian Orthodox ***"popular"*** art. That is a traditional *"folklore"* art and *not* the Royal Pharaoh's art of ancient Egypt.

The major ***characteristics*** of the Coptic icons are their extreme simplicity, and the absence of violence and cruelty in their expressions. Thus, as shown in the attached images of selected Coptic icons, round and straight lines are used; appropriate proportions are ignored; and portraits are bordering on "cartoons". The simplicity of the paintings in the Coptic icons has been interpreted as a means to expressing internal ***"calm", "peace"* and *"spirituality".*** Due to the lack of violent expressions in the Coptic icons, sometimes lions are drawn as if stripped off their savage character. Attached there are a few examples of some very beautiful, simple yet very effective Coptic icons from Egypt.

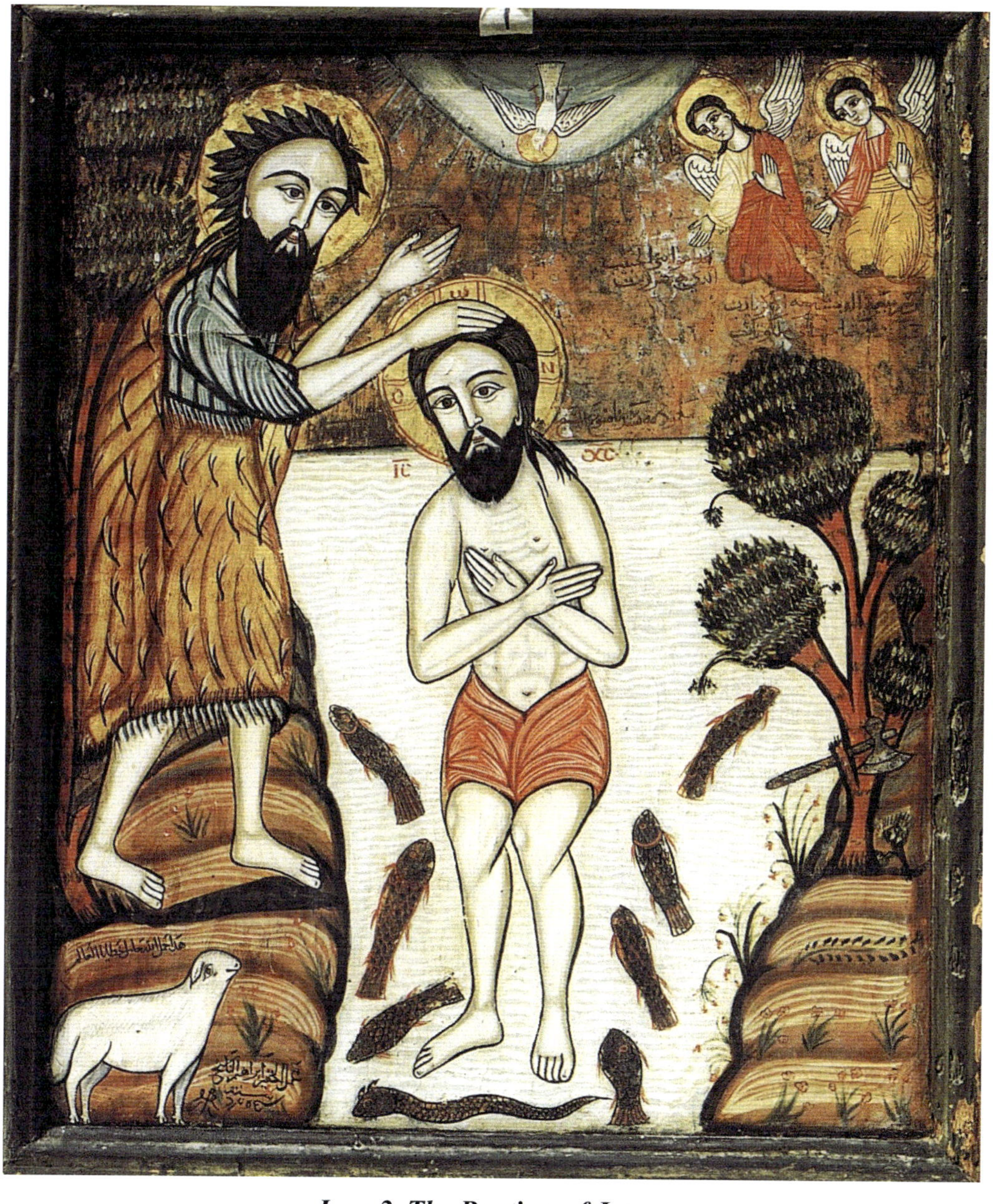

Icon 2. The Baptism of Jesus.
(An icon from Saint Mary Coptic Church in old Cairo known as Al Muallaqah (the suspended); the Church is suspended between two of the towers of the Roman Babylon Fortress.)

As regards the ***"artists"*** of the Coptic icons, we know very little about: who were they? What painting schools they belonged to? Or even what exactly the material they used and the painting methodology they adopted? The main reason is that they were "self-denied" persons; they very seldom signed their works, let alone writing about it. At most, an artist would scribble on his painting such a phrase as:" ***may God reward in Heaven those humble people who have labored in this work".*** Occasionally, however, the painter would write at the bottom of the painting something like:"***may God bless me his humble servant Ibrahim or Yuhanna (names of two famous Egyptian painters of icons from that era) for being honored to produce this work".*** Thus, painting Coptic icons was perceived as a holy job; something like praying that the artist will perform to be a good Christian; and not for fame, glory or some financial rewords.

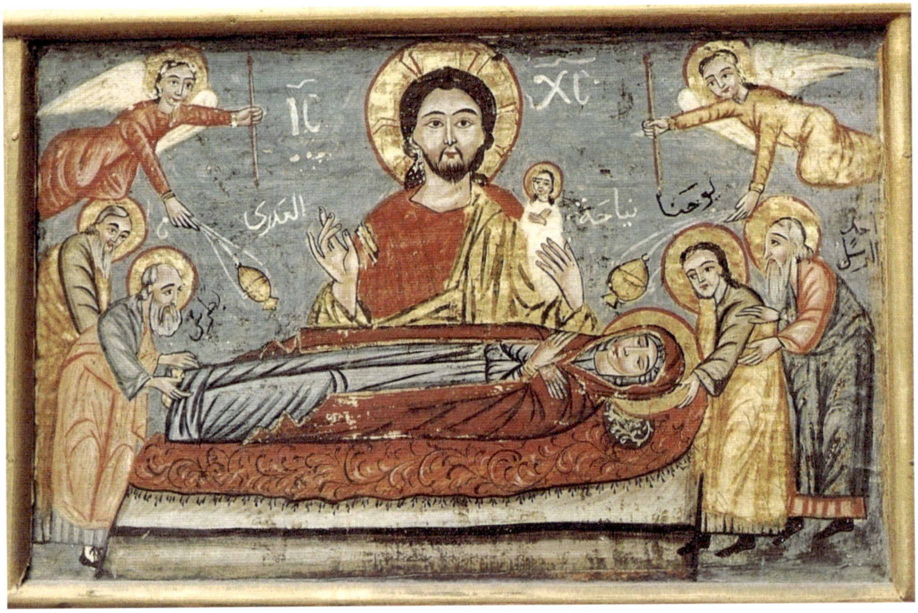

Icon 3. The Domitian of Saint Mary
(Jesus holds her soul in His arms and surrounded by angels and apostles.)

Notwithstanding the above, one can sometimes find some subtle influence of the ancient (Pharaoh's) Egyptian art on the Coptic icons. For instance, in a Coptic icon of the Archangel, He was painted carrying weighing scales as a means of weighing human actions. The scale concept finds its basis in the famous Ancient Pharaoh's Papyrious work on ***"the last judgment".***

The Coptic Churches, the Coptic museum and the Coptic monasteries in Egypt are very rich with original and very beautiful Coptic icons.

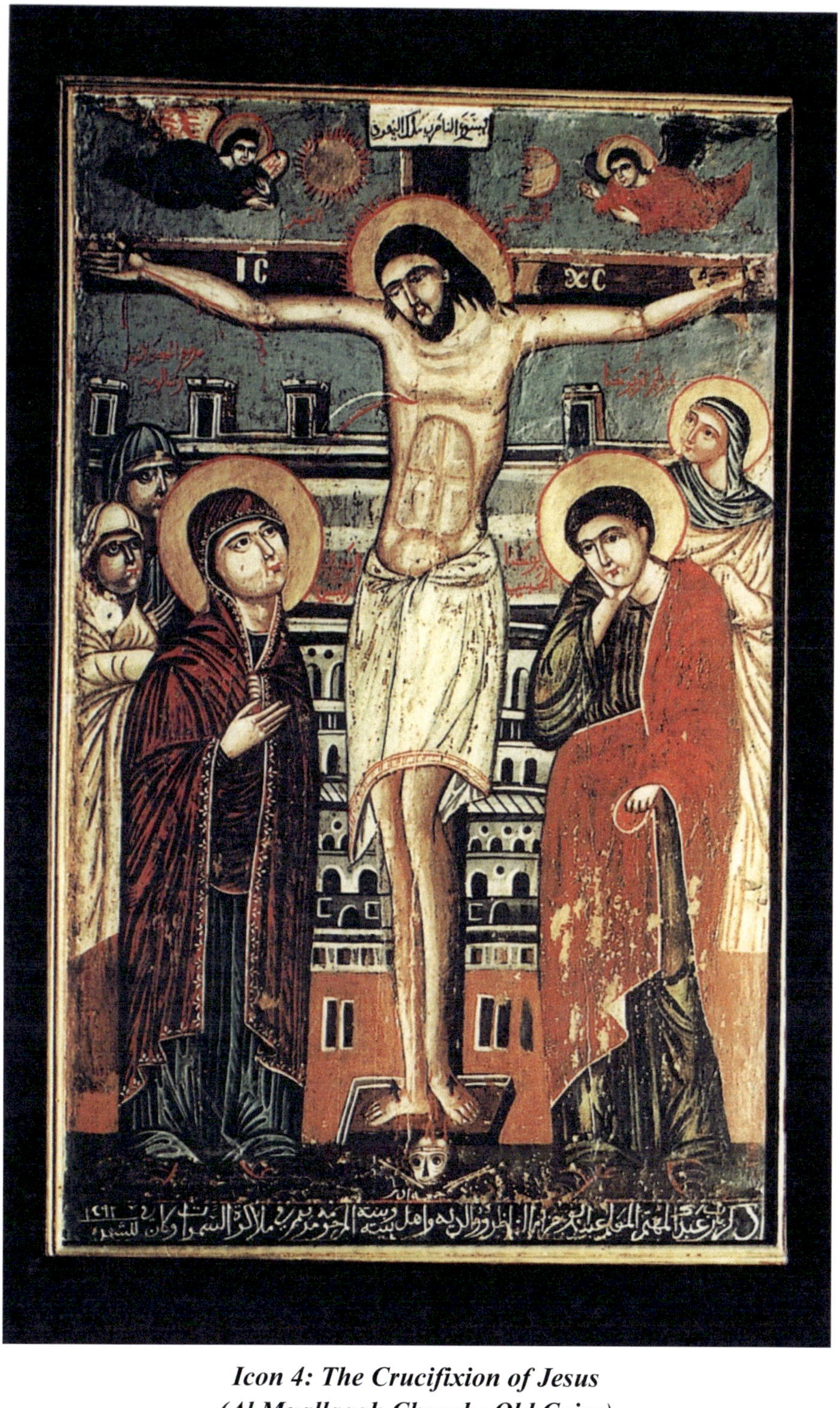

Icon 4: The Crucifixion of Jesus
(Al Muallaqah Church: Old Cairo)

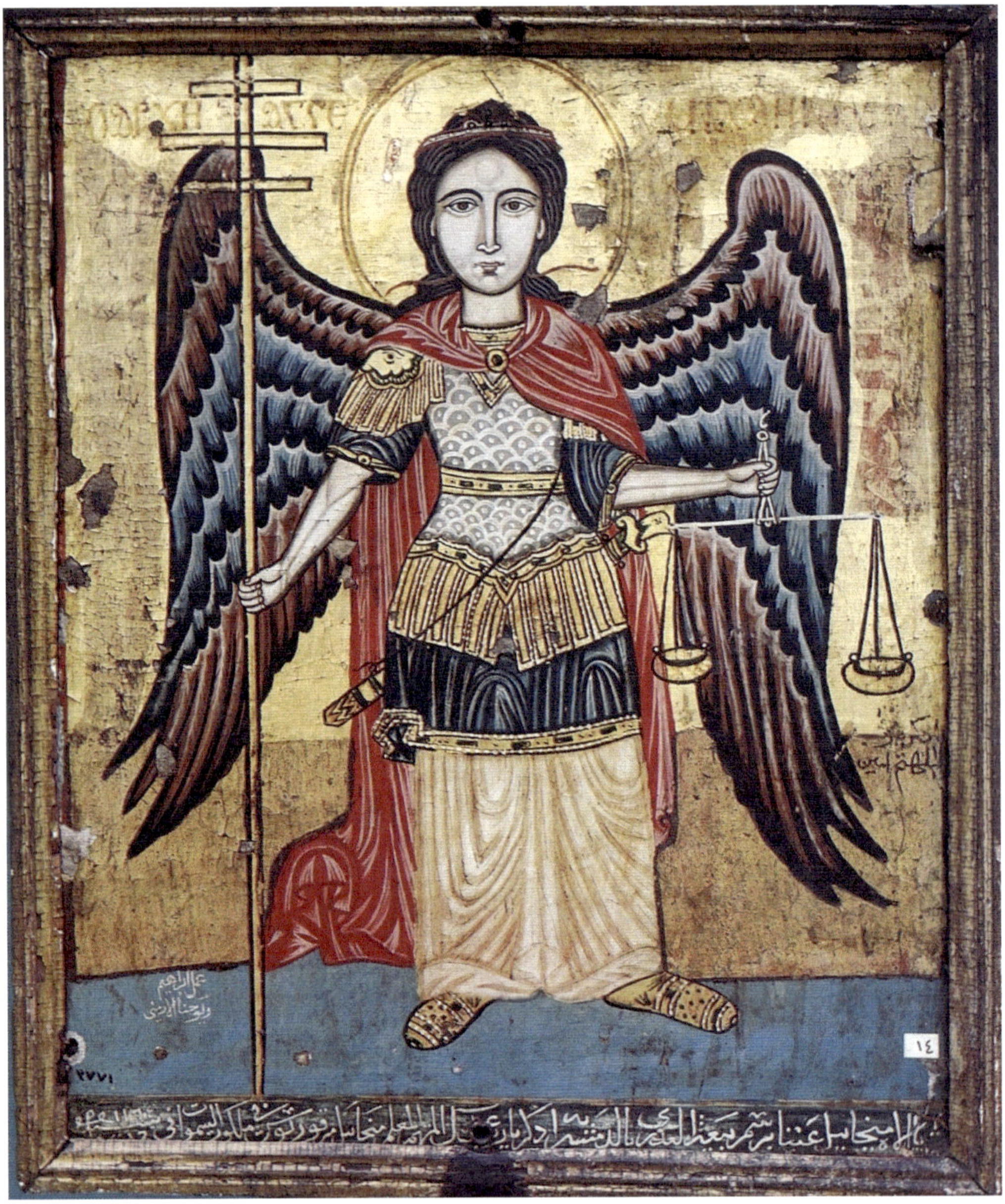

Icon 5: The Archangel Michael
(The Archangel carries the weighing scales for weighing human actions; a concept that seems to have been borrowed from the Pharaoh's "last judgment", below.)

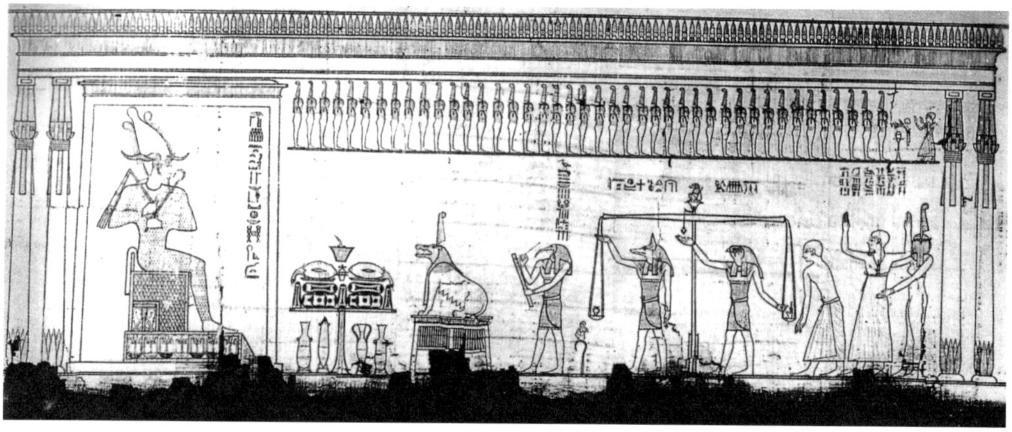

Photo 6: A Pharaoh's Painting of "The Last Judgment" over 4000 years ago, and thought to have had an influence on the Coptic art as noted above.)

III.4. Palmyra: An Ex-Empire from the East

Like people, Empires are born small, grow-up big, mature, age, and die natural death. Their monuments remain to tell us their stories.

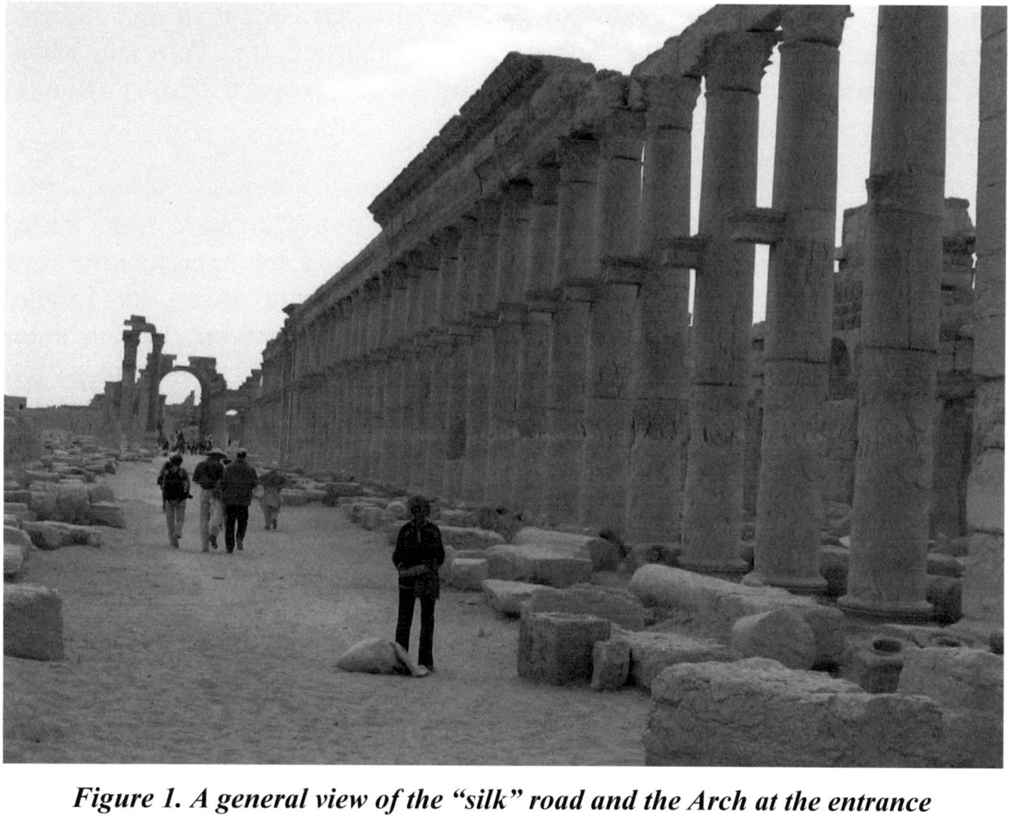

Figure 1. A general view of the "silk" road and the Arch at the entrance to the road. Statues were placed on the shelves found on the columns.

From the lush green valley of the Western Coast of Syria, you drive for about a hundred and fifty kilometers in the extended, dry and harsh Syrian Desert to reach Palmyra. Palmyra, the ***"City of Palms"***, is located in a city called ***"Tadmor"***, the ***"City of Dates"***. Palmyra is an oasis in the heart of that dry Syrian Desert. Palmyra was an Empire born over four thousand years ago, grew-up, matured, aged and died. Today its elaborate and fantastic monuments tell us the story of that Empire from the East.

Palmyra has a very impressive history. There is evidence that since earliest times (1800 BC) nomads have discovered the springs of Palmyra in the middle of the desert (see photo 2). When the Roman Emperor Trajan annexed the Kingdom of Petra (currently in Jordan) in 106 AD, the springs of Palmyra was recognized. Trade routes between Antioch (modern Antakya in southern Turkey) and Parthia (Dynasty in Iran) were established; and Palmyra, which lay in between, emerged as a caravan city of the western desert. Palmyra Empire was just born "small".

With the fall of Petra, substantive trade in such goods as: ***slaves, dried foods, spices, silk, ebony, incense, ivory, jewels and glassware*** were traded through Palmyra. The Empire then grew- up big, and for over few hundred years it became a major Empire from the East. Palmyra became an Empire encompassing a ***"hub of nations"*** from the East and the West. During those centuries of the last Millennium, Palmyra enjoyed an uninterrupted prosperity. Palmyra was a melting pot of trade, or a ***"trading centre"*** between the East and the West, particularly with the Roman Empire.

In 129 AD the Roman Emperor Hadrian visited Palmyra, after which it was briefly renamed ***Palmyra Hadriana***. During the reign of Nero, 54-68 AD, Palmyra was incorporated into the Roman Province of Syria. In 224 AD the Parthian Dynasty in Persia was replaced by the Sasanians who began to threaten the Roman rule in Syria. In 267 AD, an Arab Queen, named ***Zenobia***, rose to power and ruled Palmyra. Zenobia began to seek independence from Rome. Under her rule, Palmyrian armies captured Basra and campaigned as far away as Turkey and Egypt. Zenobia aspiration was to "divide" the Roman Empire to two parts: the Eastern part, for her to rule, and the Western part to remain under the Roman control. The Roman Emperor, Aurelian, had a different plan. He launched a campaign against Palmyra. In 272, Palmyra was defeated. Queen Zenobia was arrested by the Romans. She was taken to Rome in golden chains where she was tried and executed.

By the year 300 AD, Emperor Diocletian established a huge fortified camp in Palmyra, strengthened the walls and transformed the city into a military stronghold. Under the Sasnians rule, Persia trade via Palmyra almost seized existence causing serious decline in the Empire's wealth and prosperity. Palmyra, the Empire from the East, was thus "aging" and declining. When the Emperor Justinian refortified its walls in the sixth century, it was declared a virtual ruin. However in the Byzantine period several churches were constructed in Palmyra.

In 634 AD, Palmyra was taken by Khalid Ibn Al-Walied, one of the military leaders under the first Caliph: Abu Bakr Al-Sidek. Nevertheless, Palmyra played a minor role in the Islamic period. By the Ottoman period, the ruins of Palmyra were surrendered to the desert.

The Empire has thus "died", yet its today's fascinating and very elaborate monuments tell us a far more detailed story of that once an outstanding Empire from the East. Thus Like people, Empires are born small, grow-up big, mature, age, and die natural death. Their monuments remain to tell us their stories.

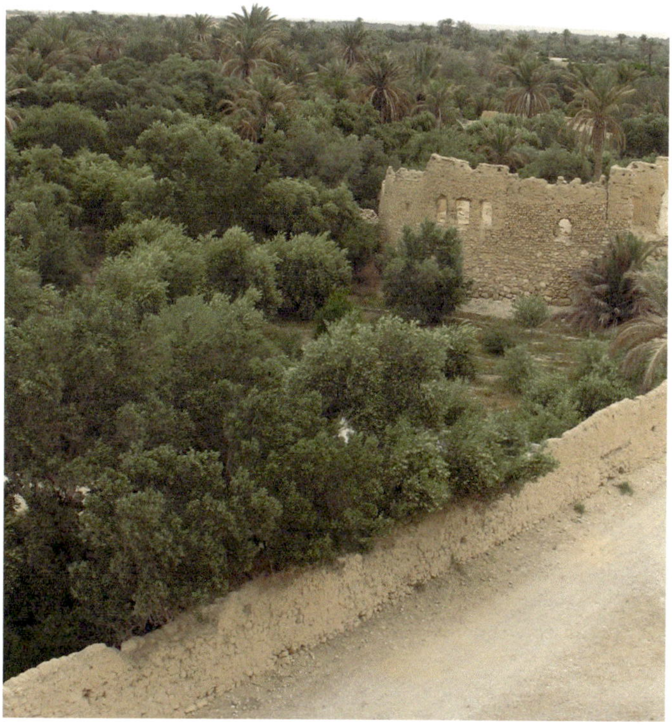

Figure 2. Palmyra's spring: until today, a green oasis in the middle of the desert.

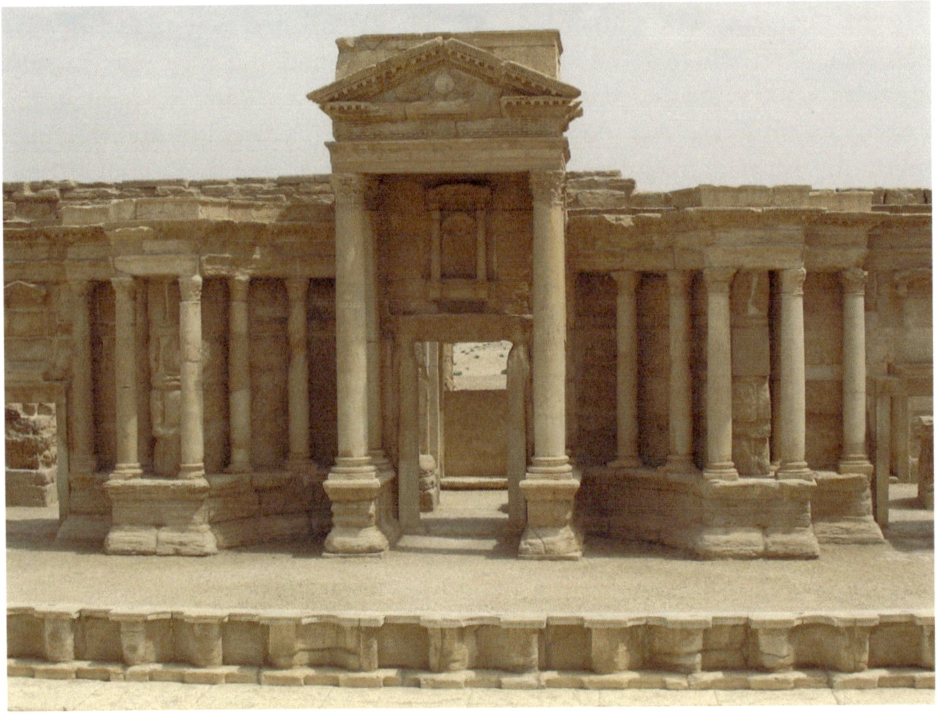

Figure 3. The Roman Theater: built in the first half of the Second century.

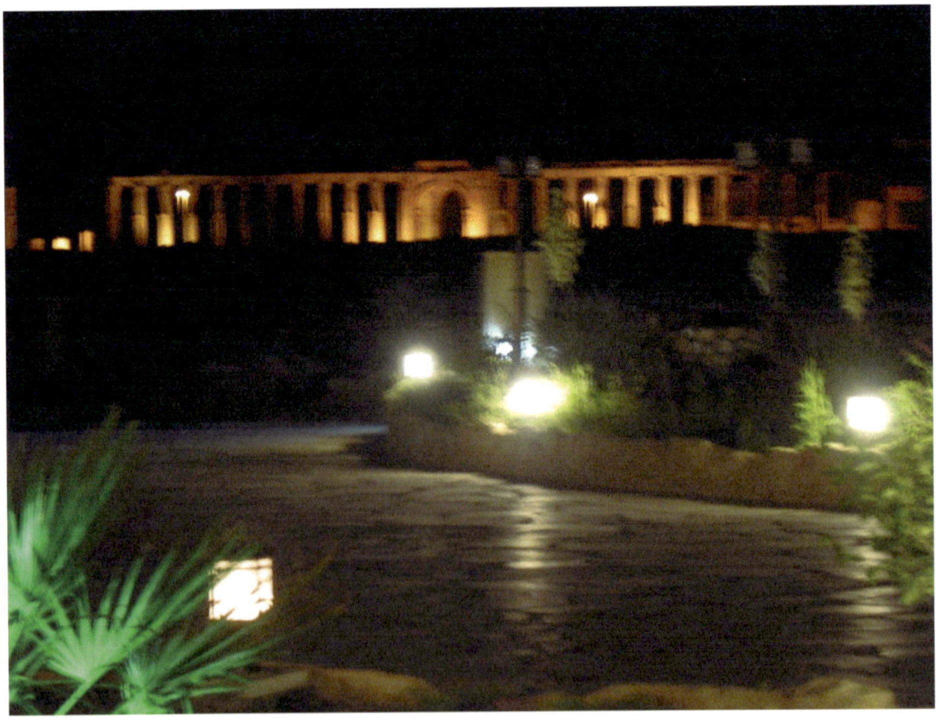

Figure 4. Palmyra at night. (A photograph by the author.)

III.5. Egyptian Obelisks in Rome

Partnership of Magnificent Civilizations

By

Dr. Lara Bishay[3]

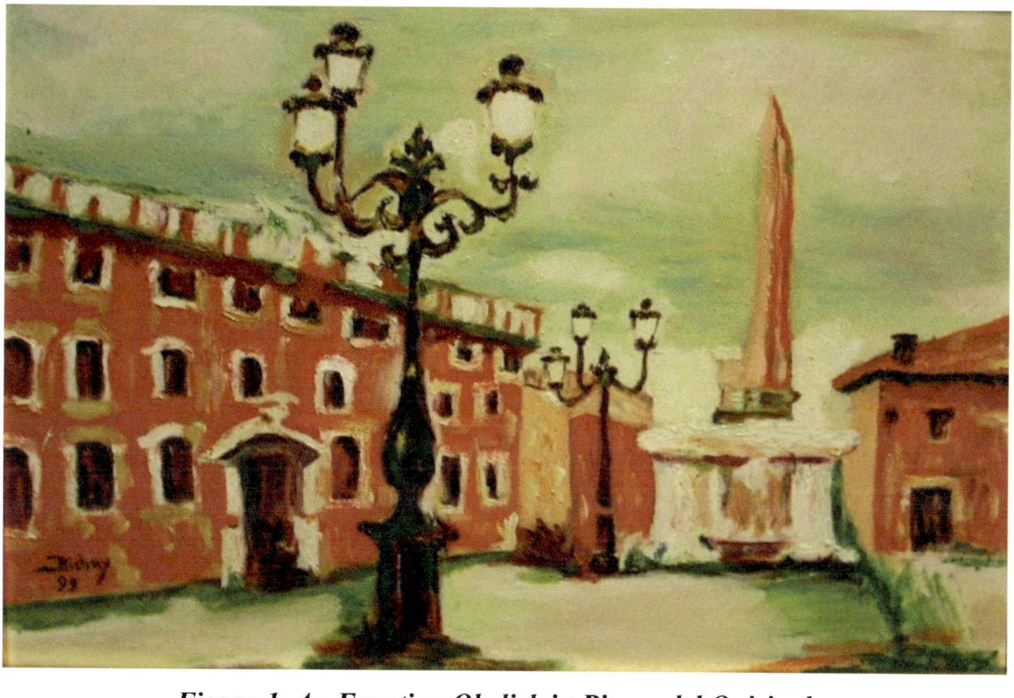

Figure 1. An Egyptian Obelisk in Piazza del Quirinale: Oil Painting by the author (oil on Canvas: 40X60 cm).

---

[3] *Dr. Lara Bishay is a pediatrician at Harvard University's Children Hospital, Boston, USA. She was born and raised in Rome. She has developed a great fascination and love for the Roman history, culture and heritage.*

A pedestrian strolling through the center of Rome will undoubtedly note the conspicuous Roman features. Among Rome's historical and beautiful locations one can find: Piazza S. Pietro, Piazza di Spagnia, Piazza Navona, Piazza del Popolo, the Pincio Park, Piazza San Giovanni in Laterano, Piazza Monte Citorio, Piazza dell' Esquilino, Piazza del Quirinale, Piazza della Minerva, and Villa Celimontana for example. While those places show no direct link with each other, they ***share one common feature***: Each of those locations contains a huge ancient ***Egyptian monument: an obelisk.*** There are 13 Egyptian obelisks spread all over Rome. Those obelisks' size, appearance, decoration, and association with the sky render them obvious, spectacular and a blatant reminder of the greatness of those responsible for their design, cut, decoration and transportation.

Obelisks in Rome were raised throughout history in three stages: (i). the people who first built and erected obelisks were the Egyptians. (ii). Roman Emperors brought some obelisks from Egypt to Rome. (iii). Roman Popes later relocated them and added crosses and Christian symbols to them (Figure 6).

Obelisks as a means of a language

During the time of the Roman Empire, obelisks were actively used as ***means of a language*** for spreading ***different messages*** to the public. Thus, the hieroglyphic depictions and the occasions for which the obelisks had been brought to Rome varied over time. The messages implied by the decision to raise an obelisk and the messages implied in their hieroglyphs both make up the general ***language of an obelisk.***

The Emperors of Rome had the means and power to use obelisks to demonstrate anything they wished, and over years adapted the obelisks to suit their own objectives. Those objectives included ***political appeals; religious motives; historical sustainability, and personal glory.***

Obelisks: Original Significance

An Egyptian obelisk is a four-sided ***single*** piece of ***granite stone,*** mostly ***cut in Aswan and Luxor.*** An obelisk ***stands*** up right, tapering as it rises. Its top is a small pyramid, called a ***pyramidion.*** The pyramidions had once been covered in ***gold*** thus reflecting light. The city of Heliopolis in Egypt had been the center of worship to the main ***Egyptian deity*[4], *Ra the Sun God***, who later became ***Ammun-Re***. Obelisks were ***sacred*** to the ***sun God***.[5] Obelisks eventually came to be used to decorate temple facades or fronts of tombs. They were used to ***celebrate the victory of a pharaoh***, who was the bodily home of a sun god. They were also erected in celebration of the pharaoh's ***jubilee,*** that is, his thirtieth anniversary of his reign and every three years after that. The ***hieroglyphic inscriptions*** on the four sides of the obelisk showed the ***glorified kings in close relationships to divinities offering*** them ***sacred symbols***. Sometimes there are illustrations of ***royal victories*** too. The hieroglyphs on the obelisk always portrayed ***a leader as a noble governor, brave and vigilant.***

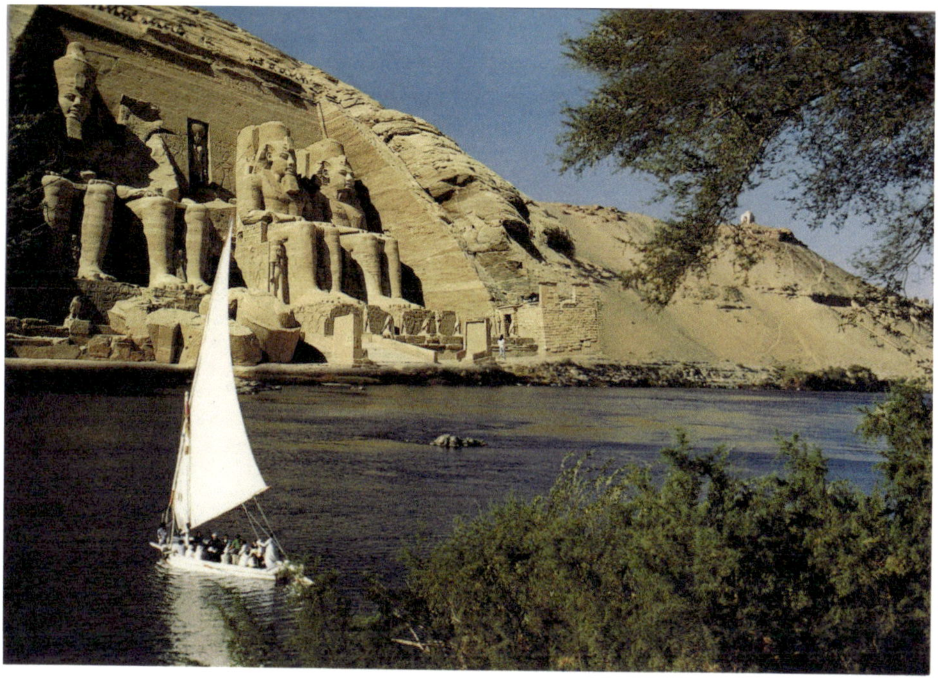

Figure 2. Aswan: where most of the Egyptian Obelisks came from.

Augustus' Obelisks: A Language of Politics

The Flaminio obelisk was the first ever brought to Rome from Egypt in 31 BC. It is ***32.77 meters tall*** (about 10 stories high), and weighs ***235 tons*** (the weight of over 200 cars). Augustus ordered it shipped from Egypt and ***initially*** set up in Circo Massimo. This was to celebrate the tenth anniversary of his conquest of Egypt at the battle of Actium when he defeated Antony and

[4] Labib Habachi, *The Obelisks of Egypt* 15

[5] Roger Lancelyn Green *Tales of Ancient Egypt* 45

Cleopatra (31BC). Later, this obelisk was moved to Piazza Del Popolo (Figure 3). Augustus did ***not*** change this obelisk in any way. It was already an artifact when he brought it to Rome. It is genuinely and authentically Egyptian.

This particular obelisk had been started by ***Seti I*** (1318-1304 BC), and was finished under ***Ramses II*** and his son ***Mineptah.*** The depictions show the Gods receiving offerings from the rulers, and glorify them for their acts of raising the obelisks. ***Seti*** is described in the hieroglyphs as ***"He who fills Heliopolis with obelisks because their rays can light the temple of Ra",*** while ***Ramses*** is portrayed as ***"He who makes numerous monuments like stars in the sky. When Ra shines he rejoices in his temple for millions of years."***[6]

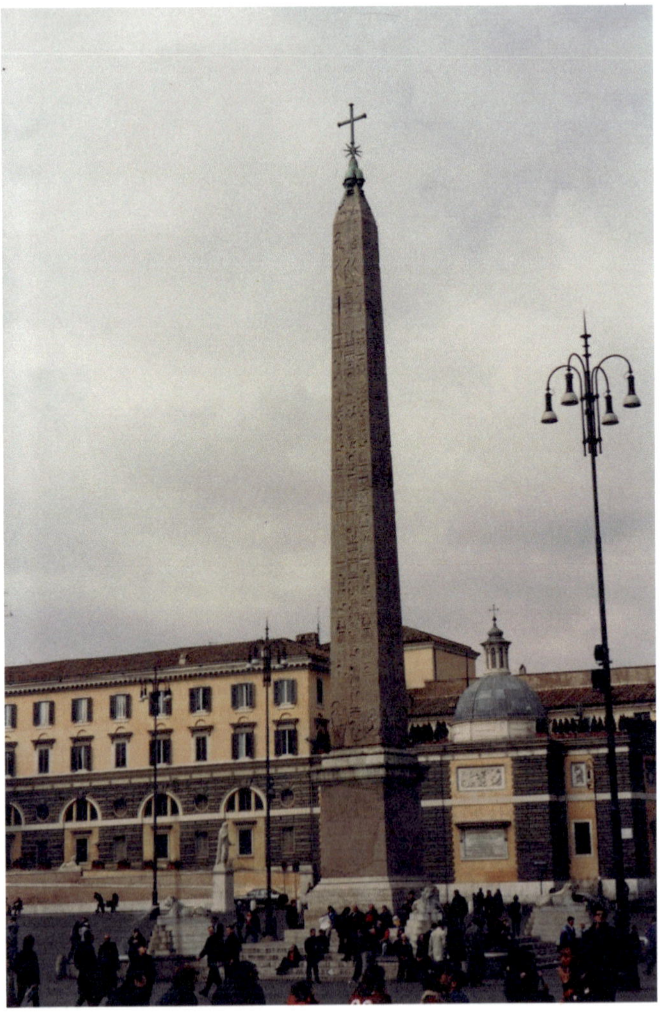

Figure 3. Egyptian Obelisk in Piazza del Popolo.

[6] Labib Habachi, *The Obelisks of Egypt* 119

The second obelisk Augustus brought to Rome is presently in Monte Citorio. Made of ***red granite***, it is ***21.79 meters high***. It reached Rome in 10 BC. Augustus had it erected in the middle of the Campus Martius. This obelisk was set on a pavement adorned with images representing the winds or zodiac signs. It seems to have served as an indicator, probably of the daily change in the southern shadow, or of the position of the constellation Horologium. That is, it was a ***celestial timekeeper.*** The inscription underneath the obelisk explicitly states: "***Augustus, son of the Deified Julius Caesar, chief priest, with emporium power, in his eleventh consulship, is the leader responsible for this obelisk".***

Domitian's Obelisk: Roman's carved hieroglyphs

Domitian brought his famous ***Piazza Navona obelisk.*** This obelisk marked a ***change*** in the manner in which obelisks were inscribed (figures 4 and 5). This obelisk is ***16.54 meters high***. The main difference between this obelisk and all of its predecessors is that Domitian's is the first to have its ***hieroglyphs carved by Romans.*** It was originally an embellishment of the entrance of the Iseum; a temple dedicated to the ***Egyptian sky goddess, Isis.***

Domitian ordered this obelisk's stone extracted from Aswan, and it was transported to Rome ***blank.*** He decided to have the ***Romans to carve the hieroglyphs themselves showing him in close relation with the divinities***.

The hieroglyphs on that obelisk describe Domitian ***receiving*** something from the gods rather than him presenting an offering to them. This was a remarkable ***novelty*** for a man to hold such a sacred position. It was also the first time that a ***Roman emperor*** actually had ***his own name in hieroglyphs engraved on an obelisk***. This obelisk was a declaration of his glory not only as a ruler, embellishing Rome with the addition of another Egyptian monument, but also his predominance in the Isis cult religion.

Bernini's Master Class of a Roman fountain and the Egyptian Obelisk in Piazza Navona

A beautiful fountain integrated with the obelisk is made by the most famous great Italian Artist: ***Gian Lorenzo Bernini***[7] ***is located in Piazza Navona.***

[7] See "Bernini: Sculptor and Architect", page 32, ATS Italia Editrice, Rome 2009.

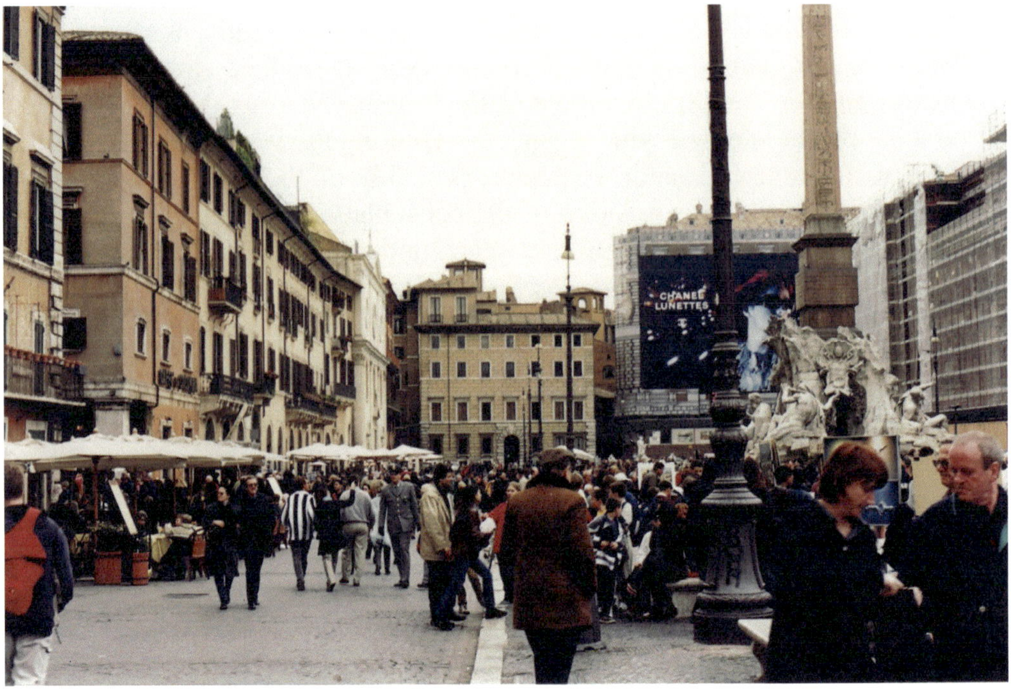

Figure 4. Egyptian Obelisk in Piazza Navona and Bernini's integrated fountain.

The Trinita dei Monti Obelisk: A Historical Record

On the top of the ***Spanish Steps*** in Piazza Di Spagnia sits the ***obelisk of Trinita dei Monti.*** What is interesting about this obelisk is that its engraved ***hieroglyphs*** are an attempt at a ***copy of the Augustus' first obelisk in Piazza dell Popolo***. After comparing particular corresponding details of the two obelisks, the Trinita dei Monti shows itself to be a "***poorer piece of workmanship".***[8] The characters are in some places facing the wrong way, which was a serious mistake in ancient Egypt because the characters were meant to point in the direction of the temple or grave they were located near to. The representations at the bottom of the obelisk have the location of the glorified king and the gods inverted. In addition the legs and arms of the depicted people are not as well defined as the original in Piazza del Popolo. The differences are striking when carefully examined.

Notwithstanding the above shortcoming of that obelisk, there are some historical messages engraved on it. The Trinita dei Monti obelisk, therefore, suggests that the language of obelisks can also serve as a ***historical medium*** as well as a political act.

[8] Boris De Rachewiltz & Anna Maria Partini *Roma Egizia* 116

Hadrian's Obelisk: A Personal Commemoration

On the Monte Pincio hill in the park amongst the park's luscious green grass and trees stands a relatively small obelisk, about ***9.2 meters*** high. It is made of granite. Hadrian, who was known for his interest in Egypt and its monuments, brought this one to Rome. This obelisk was erected for ***personal reason: Hadrian wanted to commemorate his friend, Antonius.*** Antonius died on the Nile during a trip in Egypt. He became deified both in Egypt and Rome and temples had been raised in his honor. The hieroglyphs on this obelisk are authentic Egyptian.

To conclude, great and beautiful Roman squares are glorified with Egyptian obelisks that were brought from Egypt by Roman Emperors over two thousand years ago. The Egyptian obelisks in Rome demonstrate a historical and artistic partnership of two very old, traditional and magnificent civilizations.

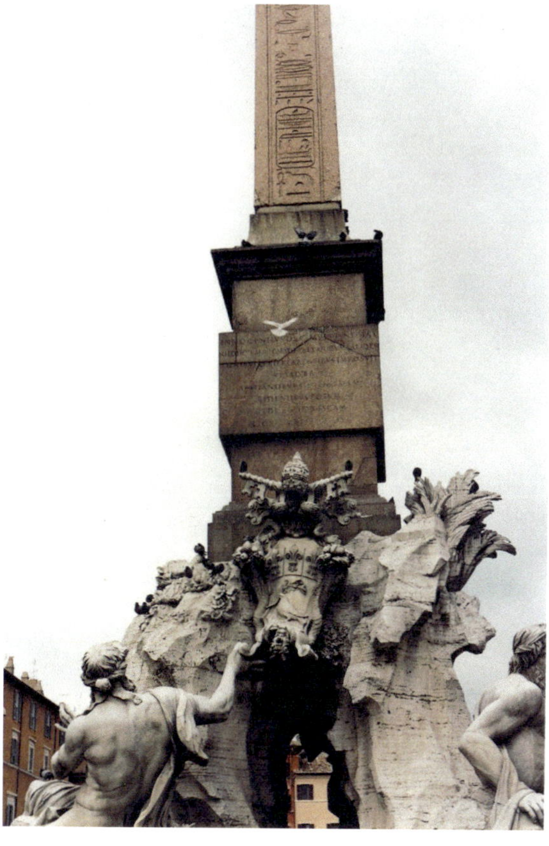

Figure 5. Bernini's Master Class of a Roman fountain integrated with the Egyptian Obelisk in Piazza Navona: Two Outstanding and everlasting Master Pieces in the World of Art.

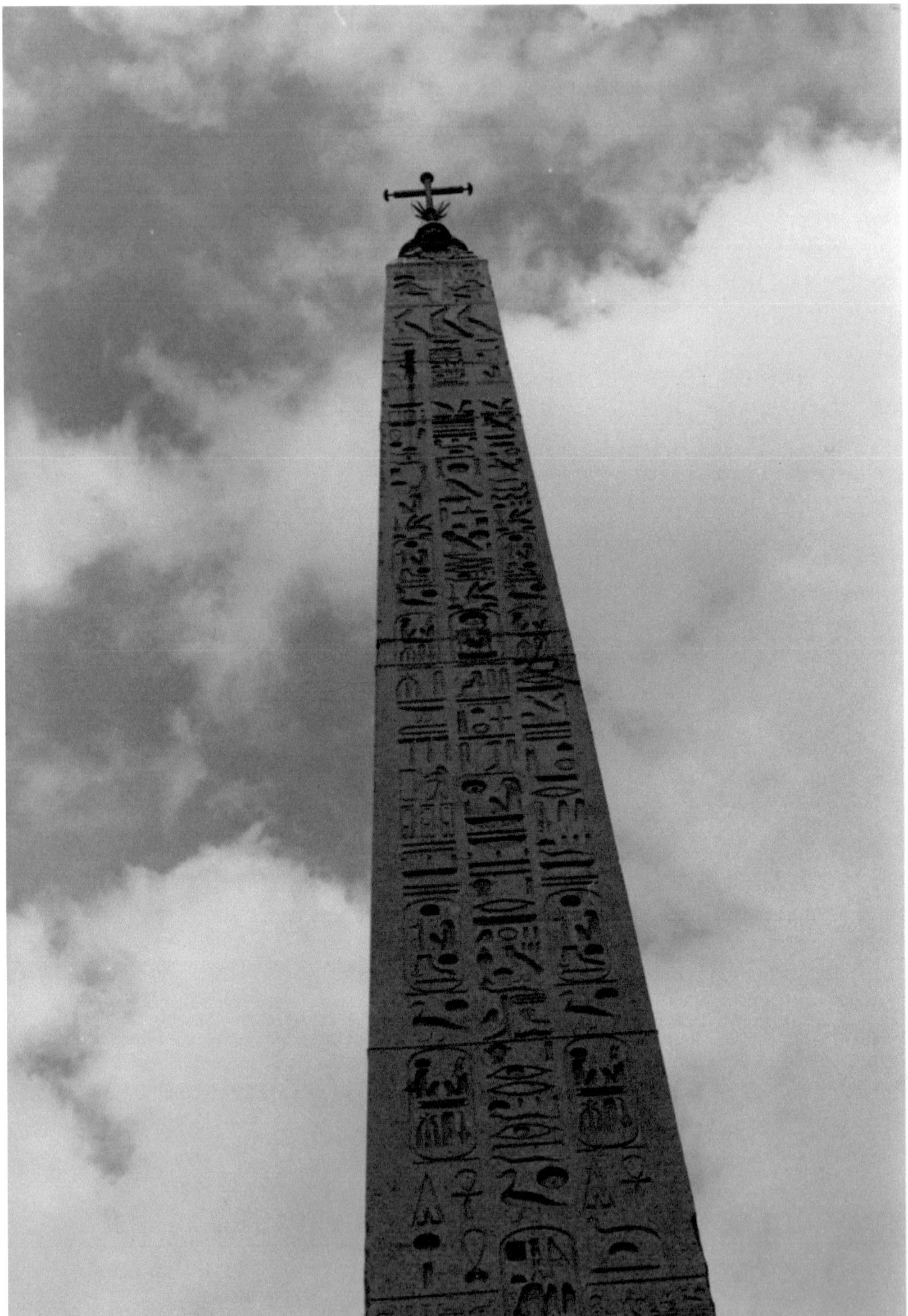

Figure 6. A Christianized Egyptian Obelisk in Rome.

III.6. Swaminarayan Akshardham

A Contemporary Marvel in Art and Architecture

It is sometimes thought that world marvels in art and architecture, such as: the Pyramids in Egypt; the Taj Mahal in India; the everlasting works of Michael Angelo and Bernini in Italy, belong to earlier centuries when societies attached higher priority to art and religion. But, Swaminarayan Akshardham Monument in New Delhi is a contemporary marvel in art and architecture that defies that thought.

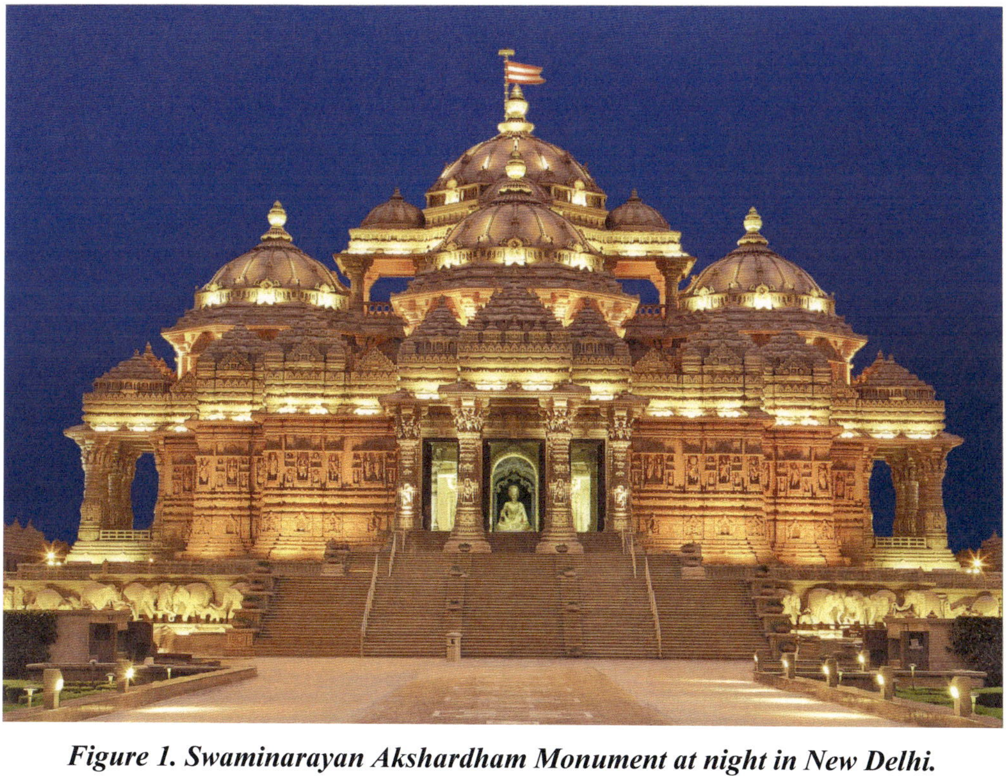

Figure 1. Swaminarayan Akshardham Monument at night in New Delhi. The Monument was built during the period from 2000 to 2005.

Introduction

India, along with a few other countries like China and Brazil, are currently labeled as ***"emerging economies".*** An emerging economy is an economy that is moving quite fast on its development path. In a relatively short time, emerging economies will assume prominent places among the world developed countries. On the move along that fast developmental track, an emerging economy faces the challenge of preserving its national heritage, culture and identity. On a recent visit to India (November 2011), the author of this article was intrigued by how India has been addressing this challenge. To this end, he visited a number of India's arts and cultural institutions to see whether the country's objective of attaching high priority to its traditional arts and culture has been maintained. One of those visited magnificent places was Swaminarayan Akshardham Monument in New Delhi. What he observed was that, India, as an emerging economy, has indeed retained its priority to art, religion and tradition.

The following are some highlights and reflections of the author's observations on the Swaminarayan Akshardham Monument: A contemporary marvel in art and architecture:

The Patronage

Swaminarayan Akshardham means the eternal abode of "***Bhagwan*** (God) ***Swaminarayan"*** (1781-1830): A torchbearer of Indian culture and spirituality. Swaminarayan Akshardham is a fabulous Monument built ***in honor and obedience*** of Swaminarayan who inspired people to lead a life of faith, love and high morality. He strongly rejected the rigid Indian caste system and the untouchability concept. He fought for women's right; and called for social justice. He preached the masses not to kill animals, to be vegetarians and to avoid all sorts of addictions. During his life on planet earth, He had over three million followers; and far more after His departure.

Contemporary Achievement

The Swaminarayan Akshardham is a relatively new construction in New Delhi. It was built during the period from the ***year 2000 to the year 2005.*** It took five years to complete. The monument is surrounded by "holy" water. That sacred water comes from 151 rivers and lakes of India creating an ambiance of a place of pilgrimage where visitors pray for spiritual progress, peace and human development.

The Structure

Swaminarayan Akshardham's Monument covers an area of about 100 acres of cultural complex. It is a manifestation of India's glorious tradition of ***art, architecture and spirituality***. It took about ***300 million man-hours*** to carve over ***300 000 stones*** to create the magnificent complex. Swaminarayan Akshardham's ***foundation pit*** covered a volume of about: ***90 meters wide, 75 meters long and 5 meters high***. The Monument was built of pink sandstone and white marble. It is about ***47 meters high*** (equivalent to 15 stories)***; 105 meters wide and 120 meters long***. It has ***234 great carved pillars; and 9 outstanding domes.*** At the center of the Monument is a ***4 meters high, Gold-plated statue of Bhagwan Swaminarayan***, (Figure 1).

Sculptures, Precious Stones and Paintings

There are more than ***20 000*** small and large ***statues*** in the Monument. All were designed and sculptured following in-depth studies and analytical work of Indian art between the 8th and 12th centuries and beyond (Figure 2). Those magnificent statues cover the walls, the columns, the ceilings and the gates of the Monument (Figure 3). Of those great statues, the above-mentioned 4-meter gold plated statue of Swaminarayan weighs 2.6 tons is placed at the centre of the Monument (Figure 1). That statue is placed on a platform that is made of white marble which is decorated with very colorful precious stones.

There are also a number of very large and colorful paintings of *Swaminarayan's* life and deeds (Figure 4). Those paintings depict His childhood encounter with friends, austerities as a child-yogi, social and spiritual services in Gujarat and his preaching of love, faith and morality.

In addition, along with Bhagwan Swaminarayan, very colorful statues of other deities are also displayed in the Monument.

All the construction design, building activities, art work, fabulous decoration, and continuing services of the Monument were (and are) performed voluntarily, with devotion, love and faith by Swaminarayan's Followers and devotees. ***The making of Swaminarayan Akshardham was, thus, an act of love, worship and thanksgiving.***

The Elephants

Elephants, in the Hindu religion, assume a prominent place. The head of the Hindu God: Ganesha is that of an elephant. The presence of elephants in a landscape implies the existence of ranges of grass in abundance symbolizing wealth, prosperity and goodness. Elephants are not only vegetarians, but also they are not vicious predators. The elephant is a symbol of endurance, patience, wisdom and power.

There are ***80 elephant stories*** from Hindu's heritage particularly between the 8th and the 12th centuries. Those stories depicted bravery, sacrifice and the continuing fight between the good and evil. Many a lesson of experience are drawn from those stories. Those stories were researched and carefully casted in stone. Each story was drawn and its ***life size*** clay model was made in Gujarat. The models were next transported to Pindvada in Rajasthan, where they were carved into ***148 stone elephants, 42 animals and 125 human figures.*** The elephants and their demonstrated stories in stone were next fixed into the outer corridors of the Monument (Figures 5 and 6). It took almost four years to create the entire setting of more than ***250 meters long and more than 3000 tons of elephant passages*** around the Monument. ***That elephant passage symbolizes mankind's love towards other livings and towards nature.***

The Landscape

The Swaminarayan Akshardham Monument's landscape covers 22 acres of beautiful gardens. There are 65 bronze statues of India's famous personalities: philosophers, intellectuals, politicians, historians and other thinkers. There are also very well maintained lawns, trees, plants and shrubs in the Monument's gardens.

In the Monument's foreground, there is a huge ***8-petalled lotus-shaped structure*** that displays prayers and words of wisdom (Figure 7). Each petal presents quotes of distinguished scientists, philosophers and statesmen from India and beyond inspiring faith in God and faith in man.

Concluding Remarks

When visiting Swaminarayan Akshardham Monument in New Delhi, the author of this article was overwhelmed and exceedingly impressed with the beauty, magnificence and the most remarkable contemporary miraculous achievement in art and architecture. He could not imagine how it was possible that such an outstanding world landmark could have been completed in only five years.

However, appreciating the enormous devotion, love and worship that were extended in building and preserving the Monument, he reached the conclusion that with love, faith and devotion, marvels can be realized; and in record times. And that was, indeed, the case with the Swaminarayan Akshardham Monument in New Delhi.

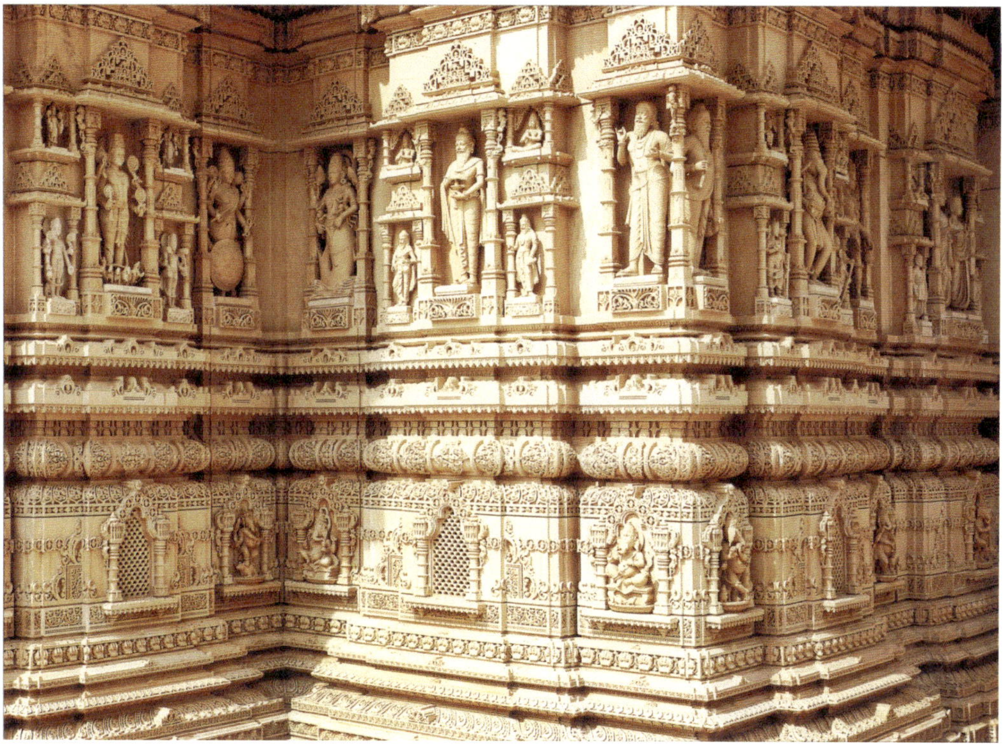

Figure 2. A side of a wall at Swaminarayan Akshardham Monument: A masterpiece in sculpture.

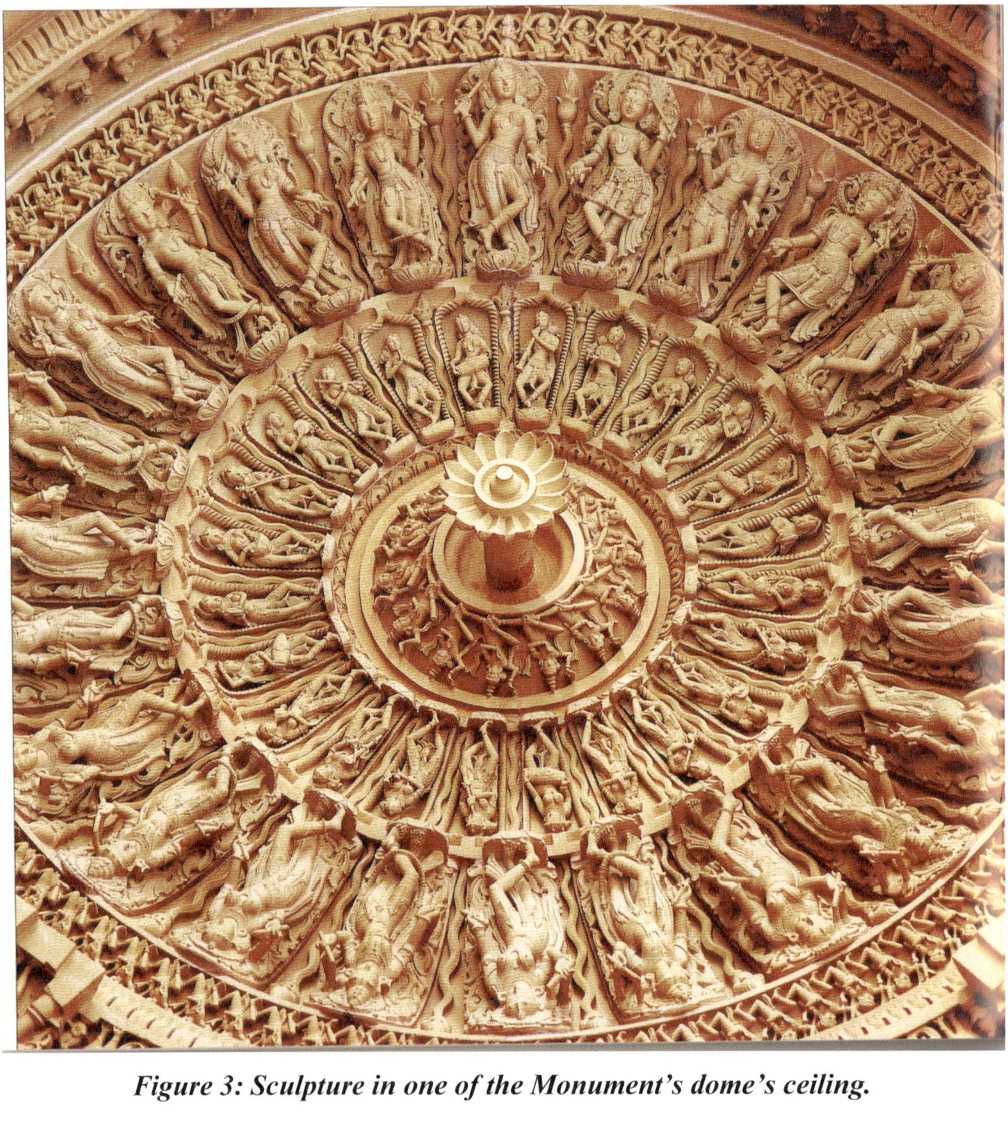

Figure 3: Sculpture in one of the Monument's dome's ceiling.

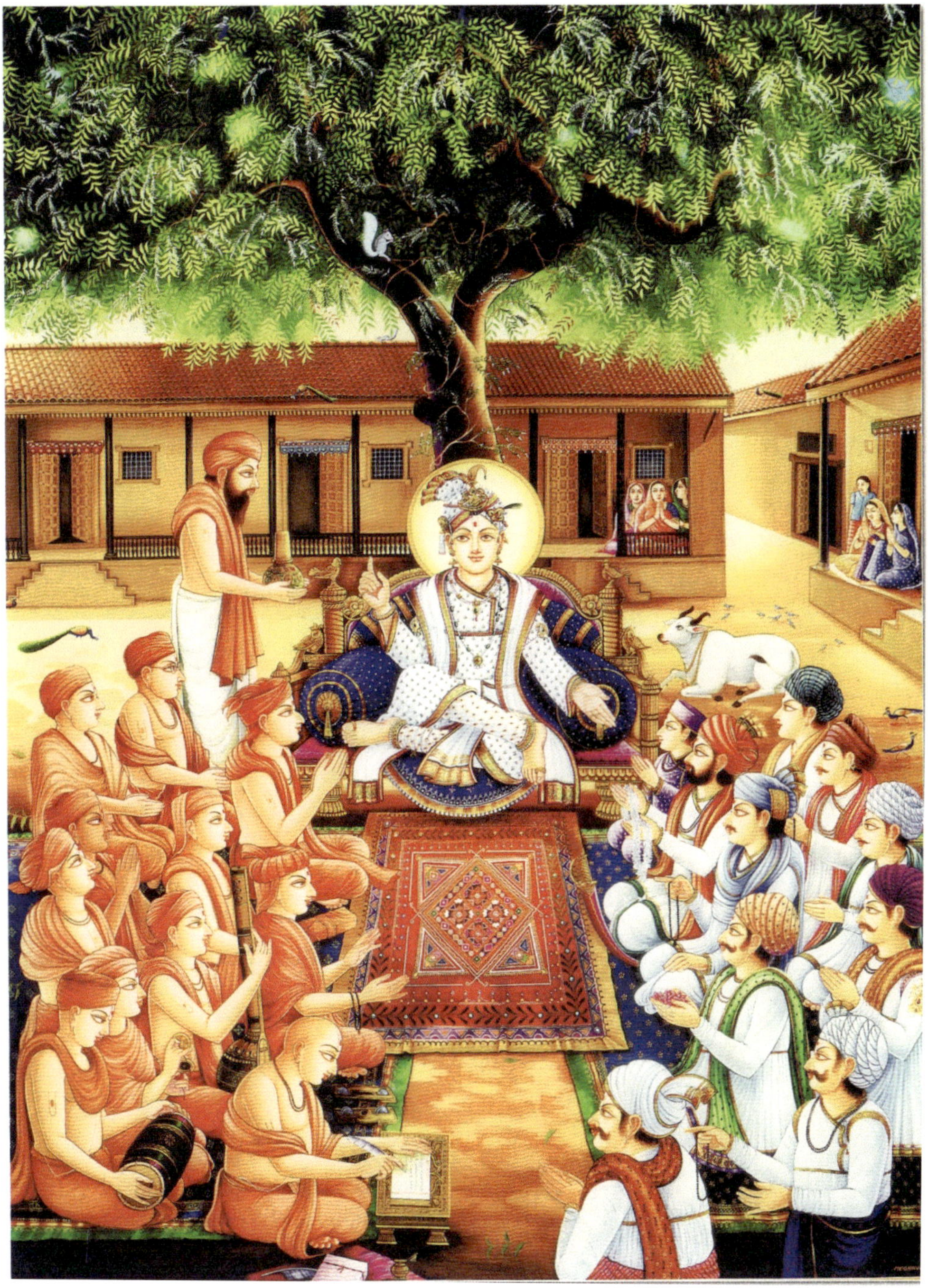

Figure 4. A painting of Bhagwan Swaminarayan preaching His message of love and faith.

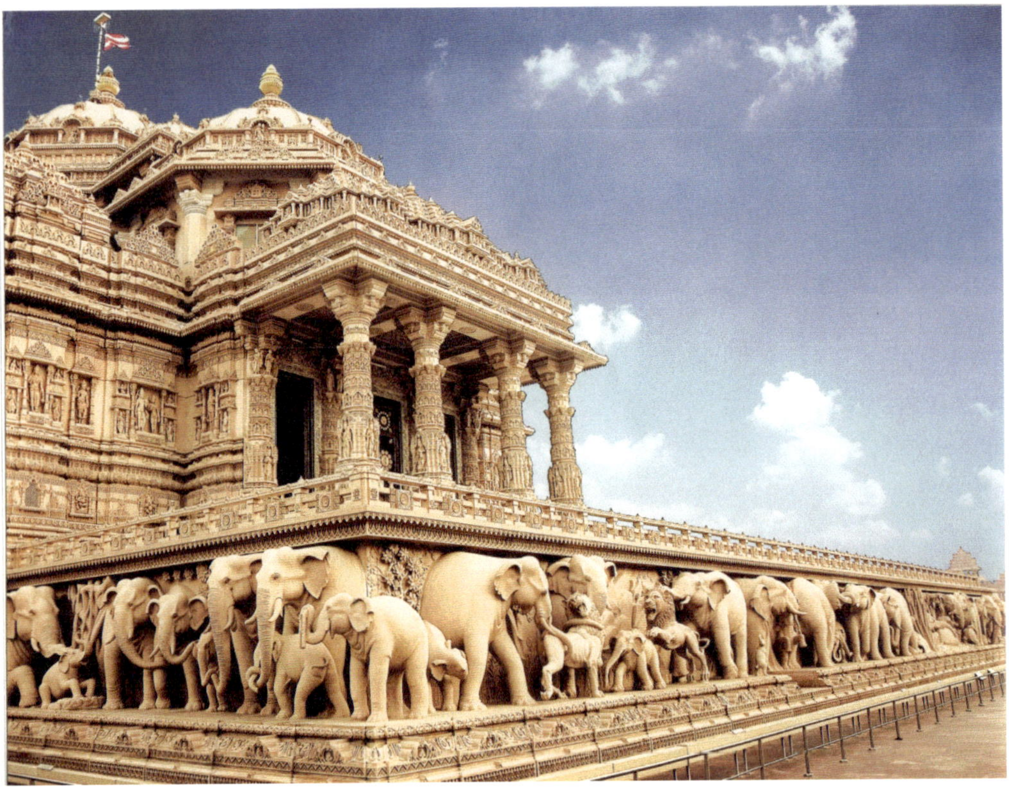

Figure 5. Part of the elephants' external corridors.

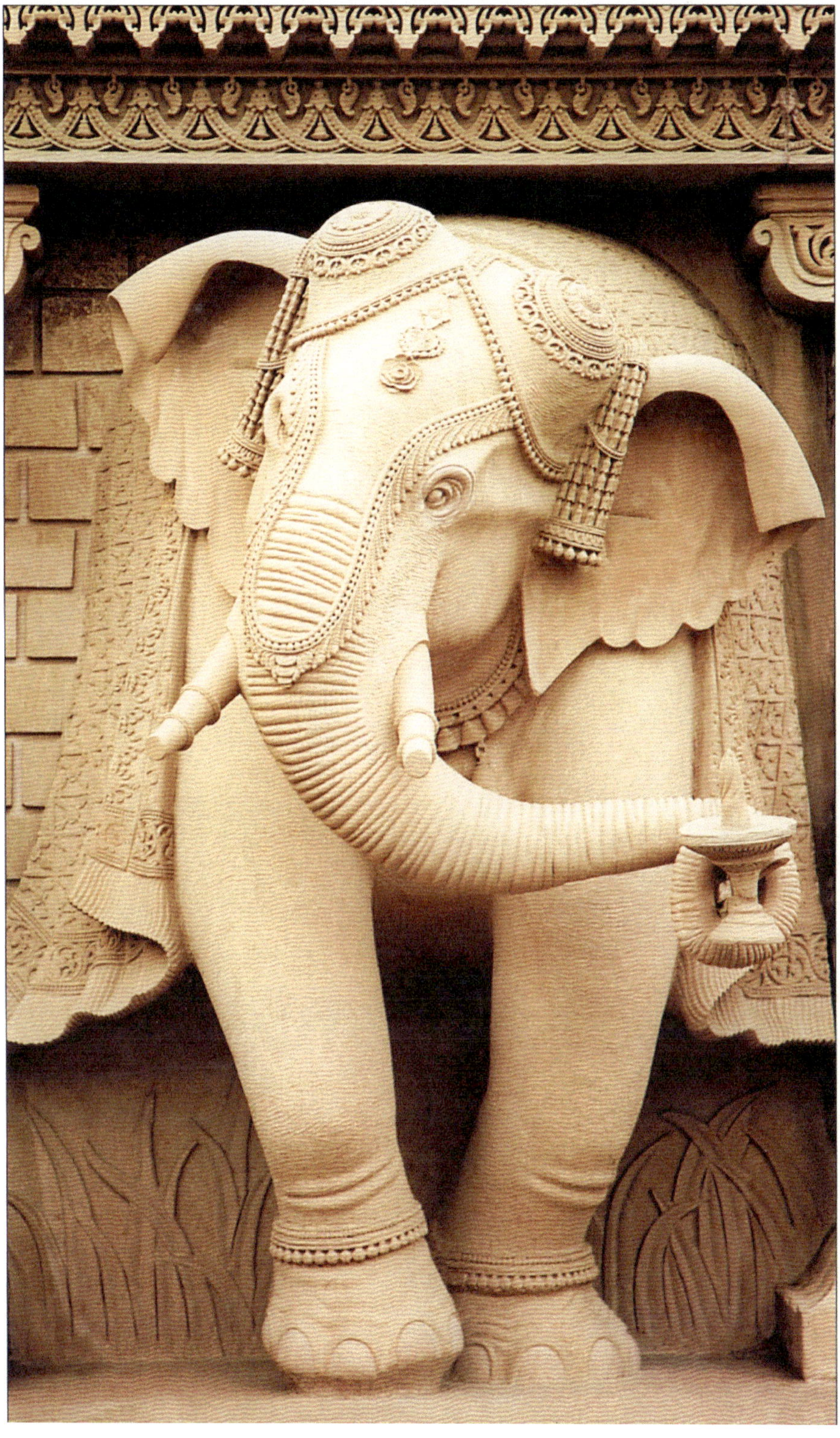

Figure6. A life-size elephant cast in stone.

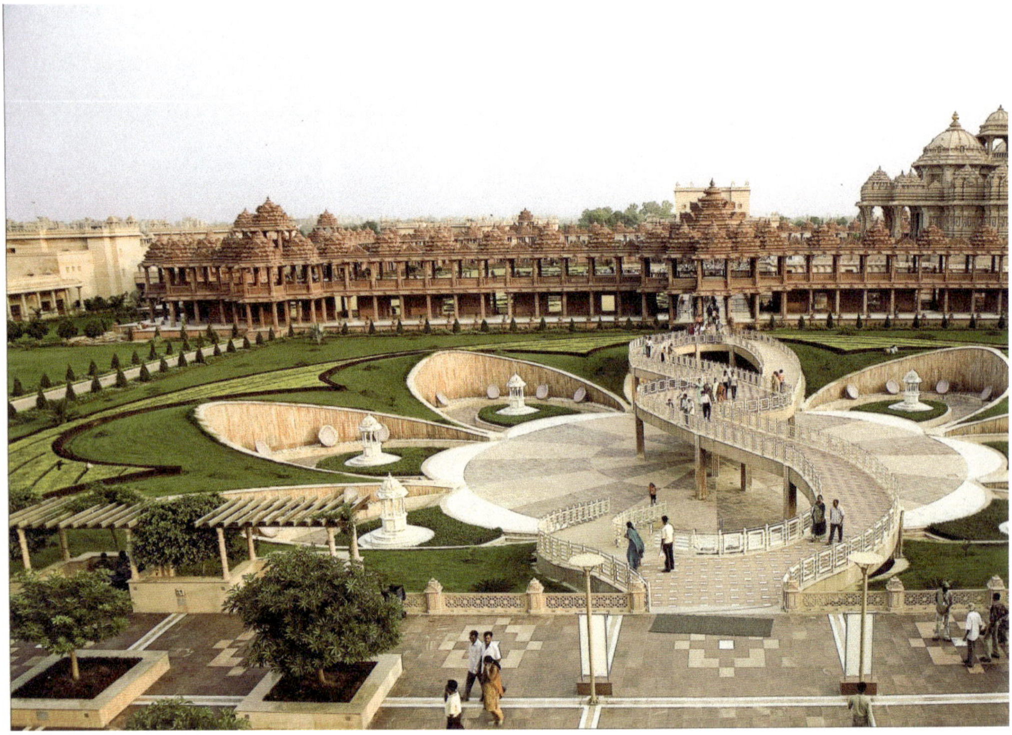

Figure 7. The 8-petalled lotus-shaped structure in the foreground of the Monument.

"What uttered from the heart alone will win the hearts of others."

Goethe

Chapter IV

On Sentimental Encounters

"How can you sing if your mouth be filled with food?
How shall your hand be raised in blessing if it is filled with gold?

Sand and Foam
Kahlil Gibran
(1926)

IV.1. A Necklace of Flowers in China

By

Dr. Lara Bishay

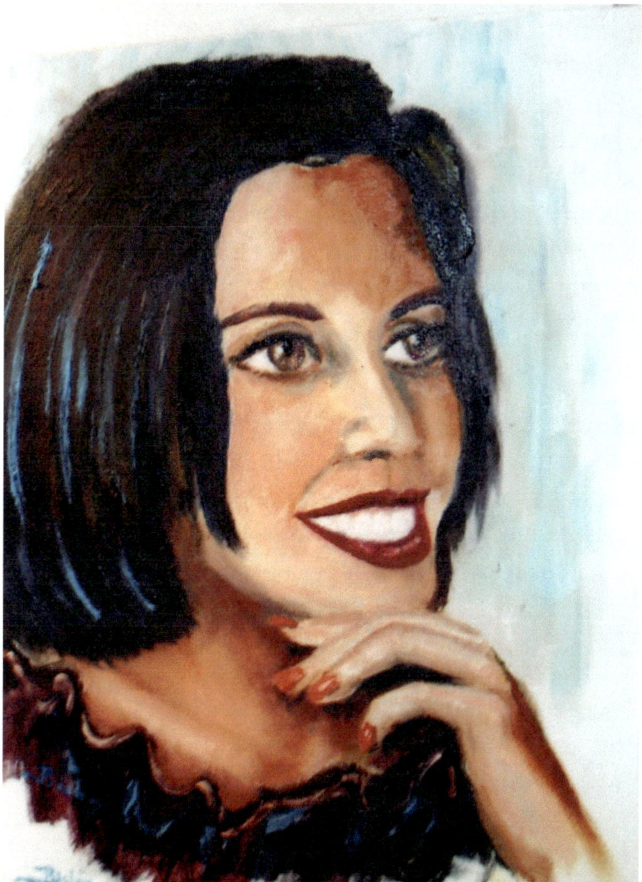

A Portrait of Dr. Lara Bishay. An oil painting by the author. (Oil on Canvas 40X30 cm.)

Through the villages we drive in our UN bus. There are puddles on the ground, dirt, and houses without doors, merely shelters against the rain. The buildings look as though they have been abandoned for years before I notice families come outside of what is their home. And yet, though the clouds of poverty and squalor, the grinning, goodness and sweetness of these destitute people surrounded me, soon to leave me shaped, influenced, inspired, and uplifted.

Little children run scattering, prancing through the fields. Their bare feet hardened and blackened by the soil, their faces glowing with the purest unblemished amiability. As we approach a tall tree, there is a strong fragrance of flowers. A little boy jumps out from behind me, and like a leopard he is swiftly climbs the tree! Back down, he bears a handful of the tree's little flowers for me. But he won't give them to me yet. His mother first sews them together. The boy takes the newly made necklace of flowers from his mother and put it around my neck laughing and talking. He wants me to remember them, and when the flowers stop smelling so nice; he will climb the tree again and get me more when I return. I want to hug them all. I want to laugh yet I feel so touched that I could cry. My face is a deep shade of scarlet in the photographs.

In these small rice fields of Sichuan, China, as I accompany a UN mission, I also discover the pure genius of the poor farmers. The rice crop grows in water about a foot high, in which fish are also cultivated. The water level drops as harvest time approaches, and the fish are collected. Rice and fish. A satiating meal by anyone's Standards. In one pond, ducks and fish can be cultivated simultaneously as well. The ducks are fed, and their excrement fertilizes the pond plants, which in turn feed the fish.

I will never forget my one-month trip to China with my father and the officers of the Food Security Project working with him in the FAO/UN. It was painful for me to see the extent of the poverty, knowing that more extreme cases exist throughout the world. I was incessantly comparing my home comforts with the conditions in front of me then. To be honest, I felt a little bit guilty about the comforts I can enjoy unlike so many people. However, the amazing and beneficent connection I saw being made between the wealthier part of the world and the local poor farmers made me want more than ever to be part of that kind of interaction.

Imagine being able to look at a person, and knowing that because of you that person's life will be better or less painful. When I received that necklace of lowers from the little boy and his mother in China, I felt as though they appreciated my presence, and were asking for it again. Imagine if I had actually done something to benefit them, like cure a disease, or help engineer more food production! I remember wishing that I had really deserved the children's smiles and invitations back. To do something constructive, that truly really benefits others around me, would complete me.

Back at school the next year, I found myself ready, with images of China in the back of my mind, to plunge into any kind of volunteer activity that would eventually aid somebody's life somewhere. These ranged from collecting money for Earthquake victims in Turkey, to flour for the starving people in Zimbabwe, to clothes for Kosovo refugees. I joined Reach Out at school, and plan to organize a service-oriented trip to the Ukraine or Albania. I cannot wait for the chances and facilities I will get in college to further allow me to be of service to the poor.

I know I am credibly fortunate to have the opportunity to receive a higher education. In my life I plan to serve less fortunate people, use my education to help them, and make intercultural connections like the one I made with that little boy and all the Chinese people I met. The China trip cemented this hope in me.

(Written by Lara when she was at high school in Rome (16 years old) and the experience stimulated her to study medicine with the objective of helping the needy. She is currently a great pediatrician.)

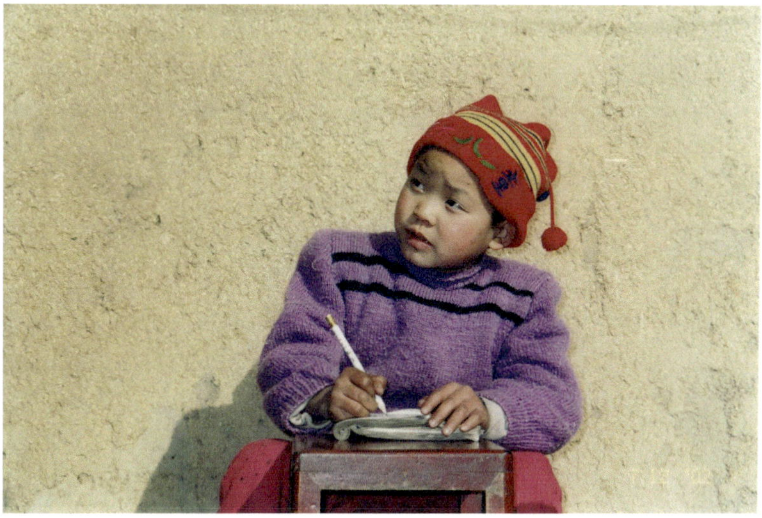

From China with love.

IV.2. "Il Lavoro Nobilita"

"...And what is it to work with love? ... It is to sow
Seeds with tenderness and reap the harvest with joy,
Even as if your beloved were to eat the fruit[9]".

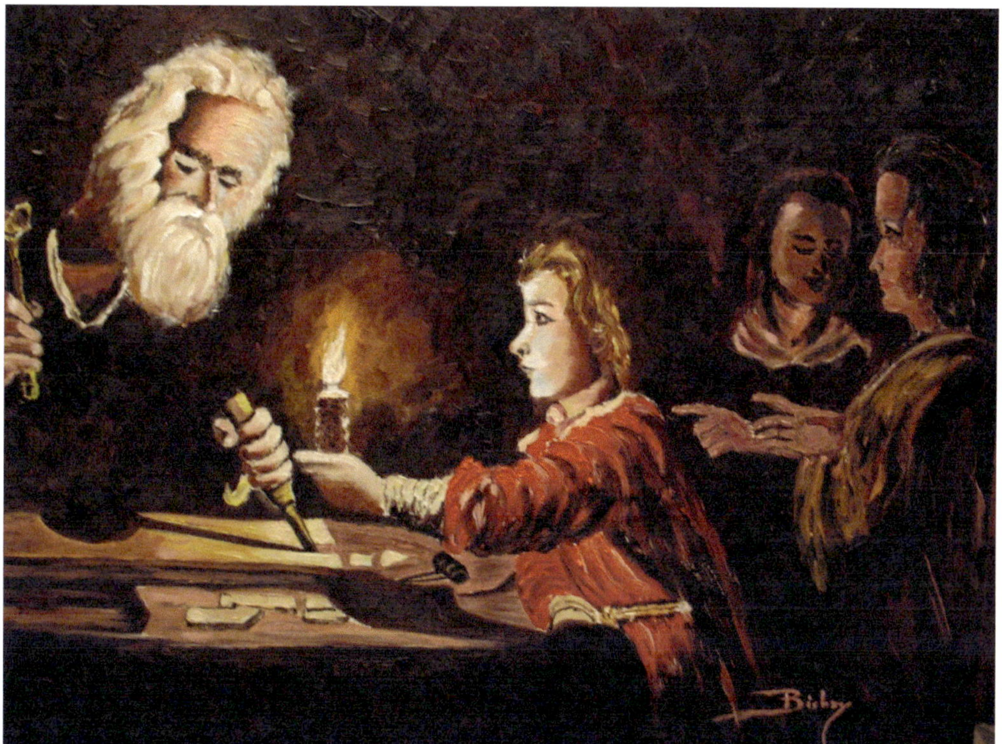

Figure 1. Carpentry labor: love made visible.
Painting by the author after the Italian Master: Bartolomeo E. Murillo, (1650).
Oil on canvas (50X70 cm)

[9] ***Kahlil Gibran, "The Prophet";*** *First published in 1923; Reprinted One hundred and thirty six times; Alfred-A-Knopf, Inc, New York, 2003.*

"Il Lavoro Nobilita" is an Old Italian proverb meaning: ***"Labor ennobles you".*** (According to Webster's Dictionary, ***ennoble*** means ***"to raise to the rank of nobility"***.) This wonderful Italian proverb applies to all kinds of (moral) labor from a head of state to a street sweeper. It is an absolute concept.

While the author of this article fully endorses the proverb and admires its inclusiveness, he, however, wishes to postulate that the proverb is actually true only when labor is performed with ***love and affection***. To appreciate this premise, let us see what Kahlil Gibran said about the nature and scope of ***working with love.***

In 1923 Gibran wrote: *"...Work (labor) is love made visible. ... You work that you may keep pace with the earth and the soul of the earth.... When you work you are a flute through whose heart the whispering of the hours turns to music.... When you work you fulfill a part of earth's further dream assigned to you when that dream was born. And in keeping yourself with labor you are in truth loving life, And to love life through labor is to be intimate with life's inmost secret."*

An outstanding example of work with love is Saint Joseph's labor in teaching carpentry to Jesus Christ (Figure 1). In the painting you can see Saint Joseph working the wood with his own hands while young Jesus Christ holds the candle for light. Observe the beauty, dignity, love and determination in the face of Saint Joseph with nobility written all over his face. And observe the young Jesus holding the candle and looking up with kindness, respect, admiration and love to the Master: Saint Joseph. During all his life on planet earth, Jesus has labored with love in teaching man how to love your enemy, in curing the sick, in raising the dead, and ultimately in sacrificing his life to safe humanity. It is in our deeds and labor in trying to follow this magnificent model that our labor can ennoble us.

Figure 3 shows another example of work with love. In the painting the young man is fetching the water for his lover. She is holding her jar and longing to her friend's gift of life.

While work with love ennobles you, work ***without love is disastrous***. Gibran further declared that:" ... *And if you cannot work with love but only with distaste, it is better that you should leave your work and sit at the gate of the temple and take alms of those who work with joy. For if you bake bread with indifference, you bake a bitter bread that feeds but half man's hunger. And if you grudge the crushing of the grapes, your grudge distils a poison in the wine.*

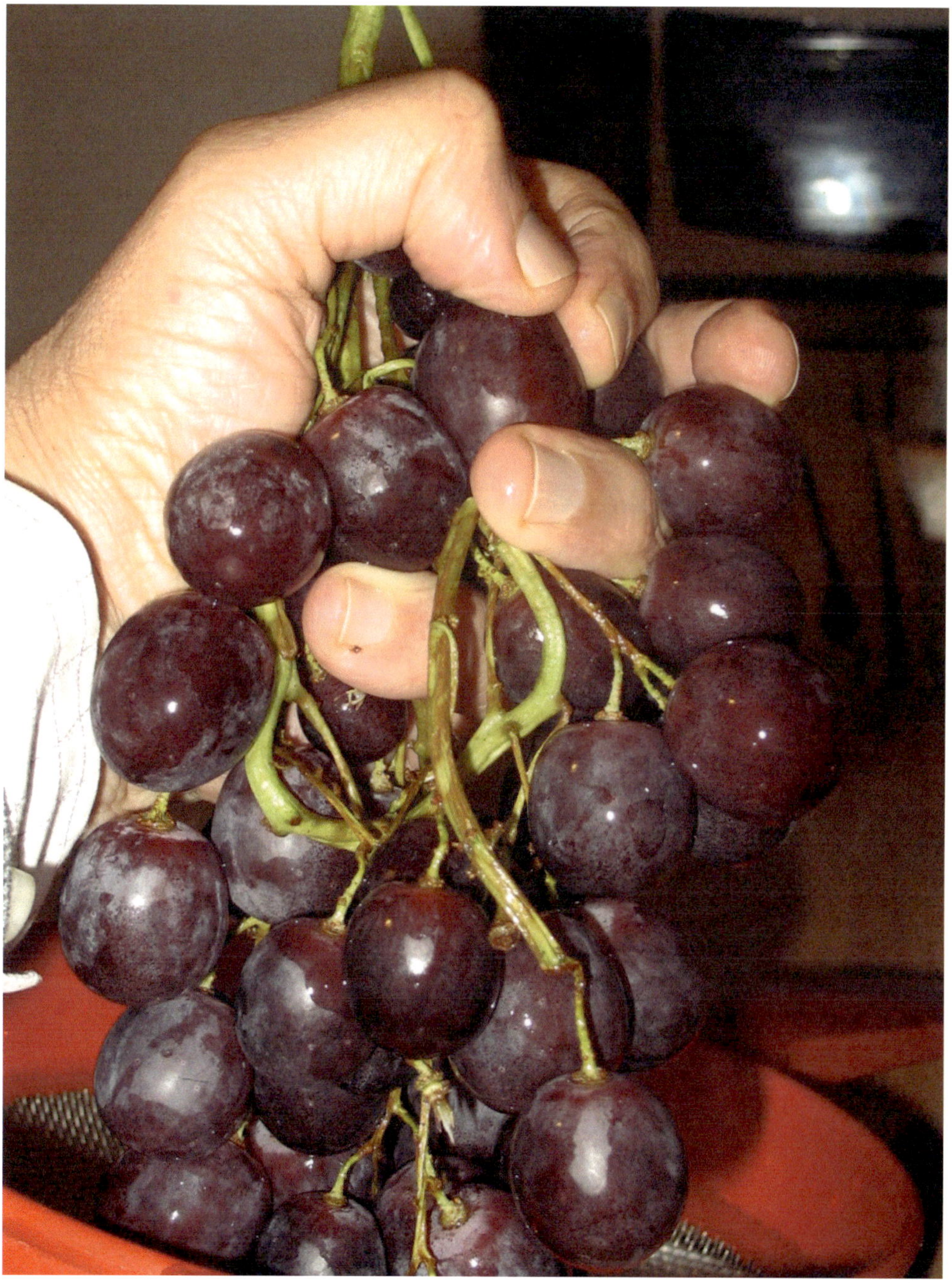

Figure 2. Even crushing grapes must be performed with love and affection lest the wine will be sour. (A Photograph by the author.)

And if you sing though as angels, and love not the singing, you muffle man's ears to the voices of the day and the voices of the night" (Ibid p.25).

While ***all*** works must be performed with love and thanksgiving, those of us who work/worked with FAO are especially privileged. Our mission has been to serve the needy in member countries (in relation to developing their food and agricultural sectors) from the depth of our hearts. Working with FAO gives us, indeed, the opportunity to appreciate, help and directly contribute to assessing the needs and achieving the aspirations of those countries. No work is nobler than that.

Member countries trust us, and open their hearts and reveal their secrets to us in order for us to labor with them in appreciating their challenges and assisting in finding pragmatic solutions to their technical and socio-economic problems.

Thus, it is they, the needy and our friends in member countries, who bestow on us the ***honor*** to help them. It is they, who give us the opportunity to give not from our possessions, but from ourselves, and that is when we truly give. It is they who treat us as their ***"healers"*** in assisting them in trying to overcome hunger, improve food security and in putting them on the right track towards sustainable agricultural and rural development. It is they who bring the best in us through granting us the honor and the privilege to befriend them, to love them and help them on the road towards development and prosperity. It is they who, indeed, make "***our labor ennobles us***"; or make ***"Il lavoro ci nobilita".***

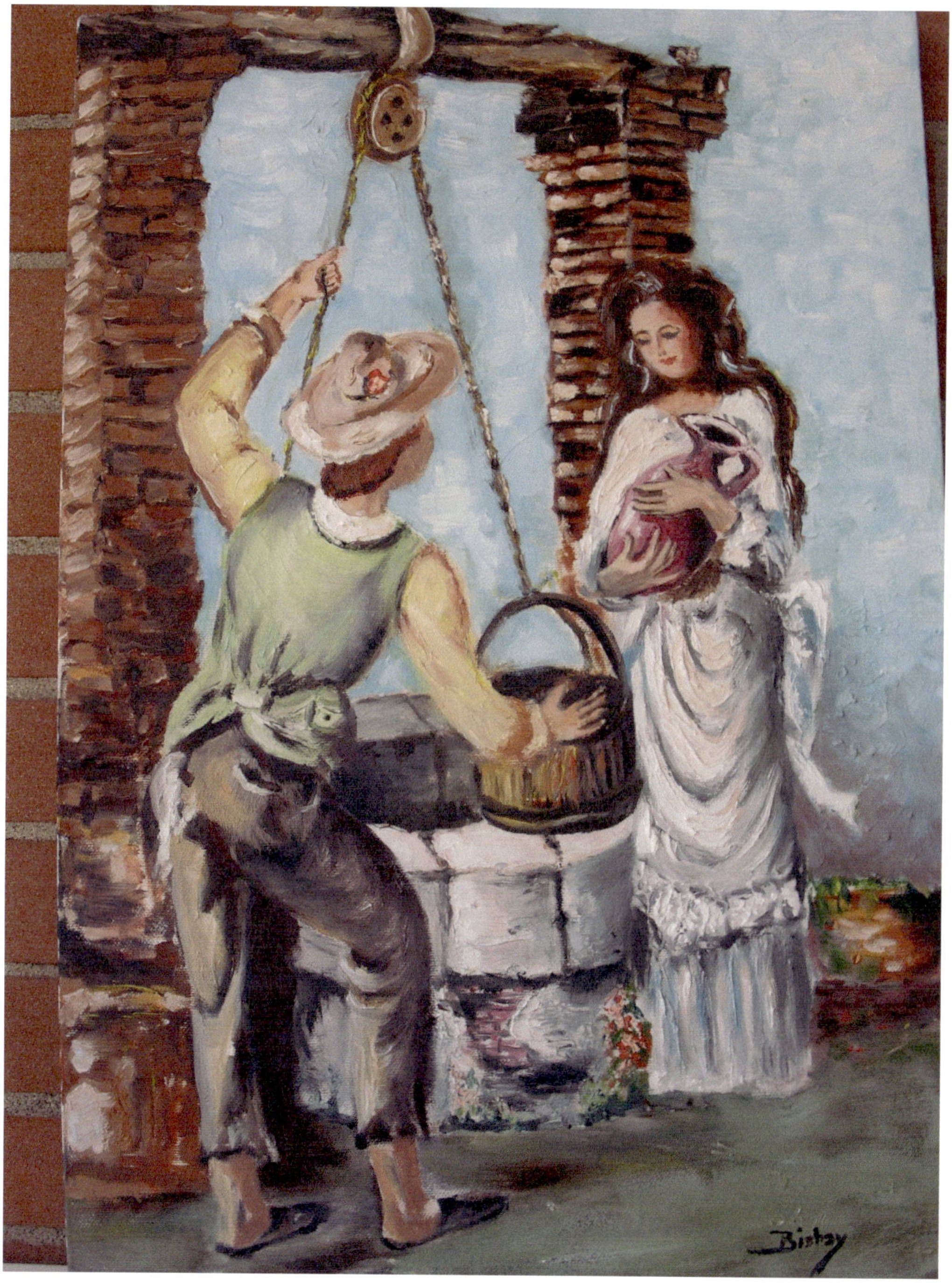

Figure 3. Working with Love: Longing for the gift of Life. Oil painting by the author (oil on canvas 60X40 cm.)

"If you truly love nature, you will find beauty everywhere."

Vincent van Gogh.

Chapter V

On The Art of Aging

"When we age with grace, and we all can and should,
we shall have faith and real strength."

The Author

V.1. Vintage Beauty

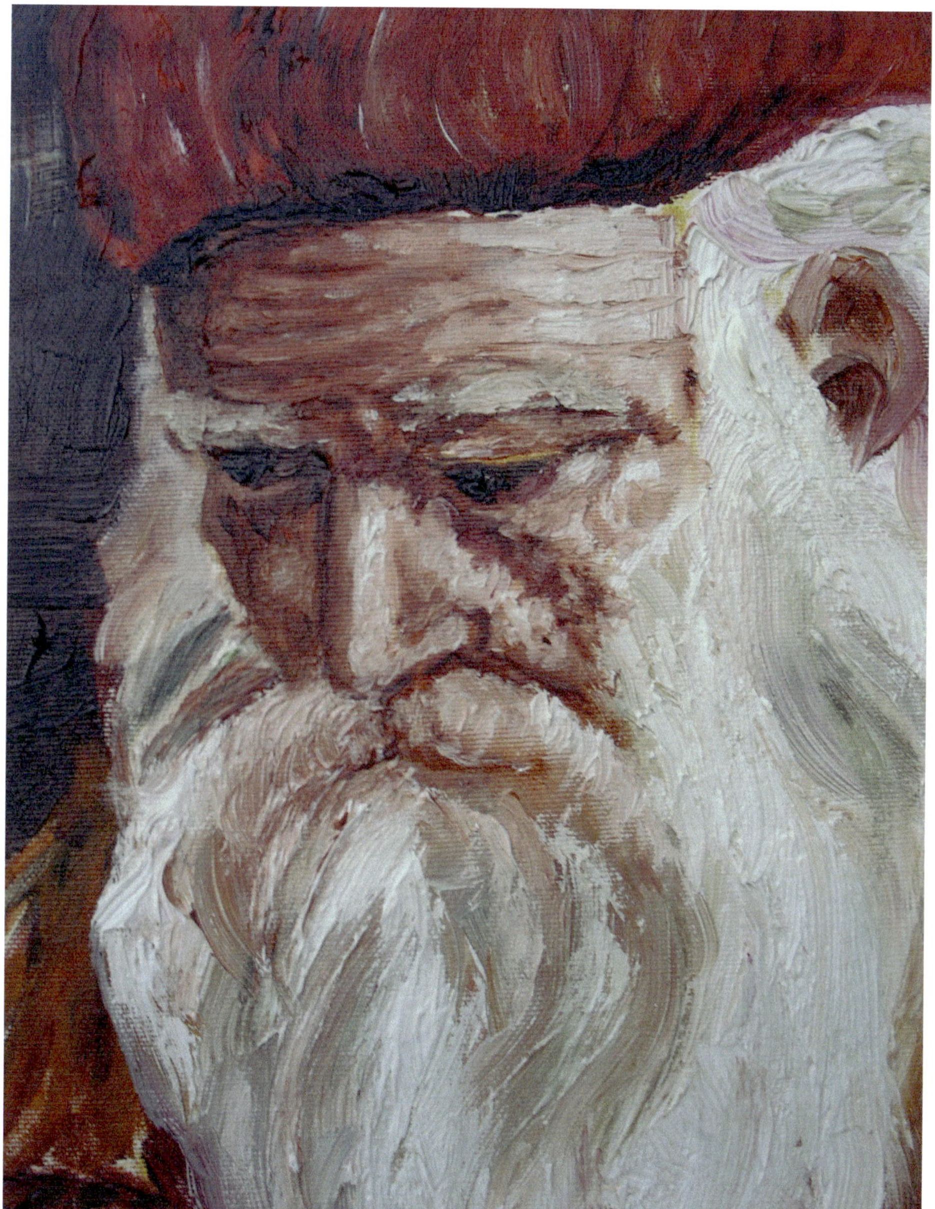

Portrait 1: "Embracing the Past with Remembrance and the Future with Longing"
Painting by the author, Oil on Canvas (60X40 cm)

One day we shall all be blessed with getting old. For ***getting old is a blessing***. When we age ***with grace***, and we all can and should, we shall have faith and real strength. We shall realize that our power to love is boundless. We shall have deep and solid personal relationships. We shall realize that career, important as it may be, does not define our identity. We shall gratefully accept the fact that some of our goals will remain unaccomplished. We shall acquire wisdom to appreciate that yesterday is but today's memory and tomorrow is today's dream. We shall embrace the past with remembrance and the future with longing. We shall develop beautiful thoughts such as those of ***beauty*** itself as perceived by Kahlil Gibran: that beauty is us. He said: ***"Beauty is life when life unveils her holy face. But you are life and you are the veil. Beauty is eternity gazing at itself in a mirror. But you are eternity and you are the mirror"***[10].

Art has captured ***vintage beauty.*** In the attached paintings by the author, we can appreciate such beautiful attributes of the old.

Portrait 1: "Embracing the Past with Remembrance and the Future with Longing"

We can see in the eyes, face and even the beard of our friend: the old man, the wisdom and deep thinking. Those are among the great attributes of getting old. When we are old we have the pleasure and the time to ponder and to think about subjects of our own liking. A pleasure that the youth often overlook for being preoccupied with career development and material achievements. Our friend in this painting is perhaps ***worried*** about wars, death and destructions in many a part in our world. He is perhaps very worried about man-made conflicts and wars, and very concerned about natural disasters. He is especially unhappy about the former more than the latter. For the first are avoidable with true love and brotherhood, but the latter are not. Perhaps his heart is bleeding especially for the man-made crises, yet he is grateful for the efforts to respond to such nature's disasters as the Asian tsunami and the Haiti's earthquake. In his wisdom, perhaps he is recalling ***lessons of experience*** from his long and rich life to draw upon for solving today's problems, and for achieving the aspiration of a better tomorrow. Our friend is, thus, embracing the past with remembrance and the future with longing.

[10] Kahlil Gibran, "The Prophet", originally published in 1923, and later republished by Alfred. A Knopf, New York, in 2003.

Portrait 2:"Deep Meditation"

Here our old friend puts his hands on his head for concentration. When we age we have so many stories and life experiences that make meditation and reflections on life a very rich and fulfilling experience. Meditation cleanses our soul and mind. Meditation opens the spring of love in us. Meditation deepens our appreciation of our fellow brothers and sisters. Meditation enlightens our souls with love and kindness. And positive and spirtual meditation preaches love and peace and rejects hate, wars and destruction. Our friend in this portrait is perhaps unhappy as he longs for a world with more love, peace and prosperty. Perhaps he is praying that all of us love each other as dear brothers and sisters. And, are not we all brothers and sisters? Note the deep emotins in our old firiend's face and his body language while he is in a deep meditation.

Portrait 3: "Saint Joseph: The Carpenter"

This is a painting by the auther after Bartolomeo E. Murillo (1650): one of the old Italiam Masters. Here Saint Joseph is teaching Young Jesus the carpentary trade. See the beauty, the calm yet the skill, determination and concentration of old Saint Joseph. See also how Jesus Christ holds the candle and looks up with love and admiaration to the Master: old Saint Joseph. When we age we become teachers and the youth become our students. We teach them with love, faith and skill. When we age we become ***old, resourseful and beautiful.*** Getting old is, indeed, a blessing.

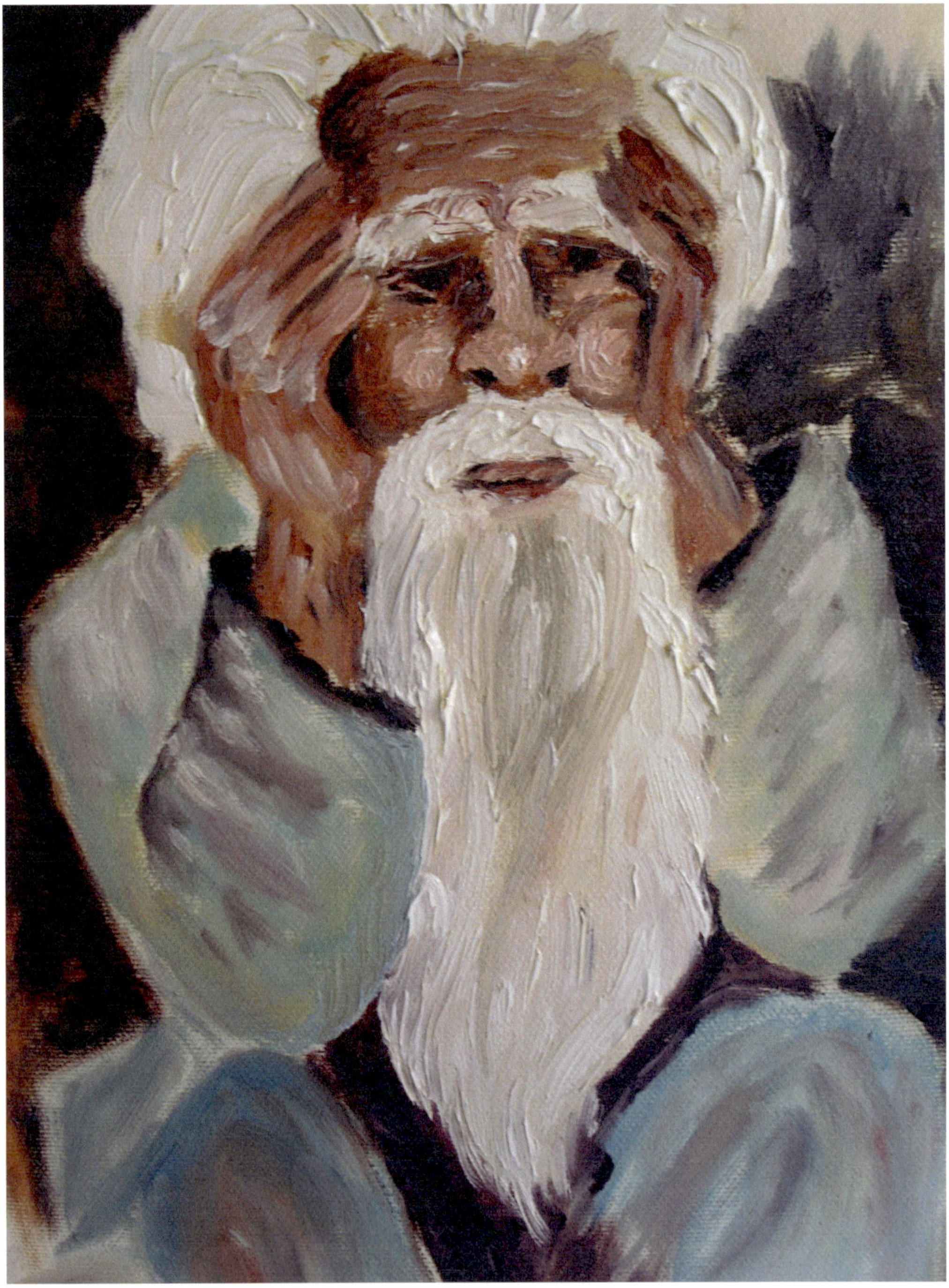

Portrait 2: "Deep Meditation"
Painting by the author, Oil on Canvas (60X40 cm)

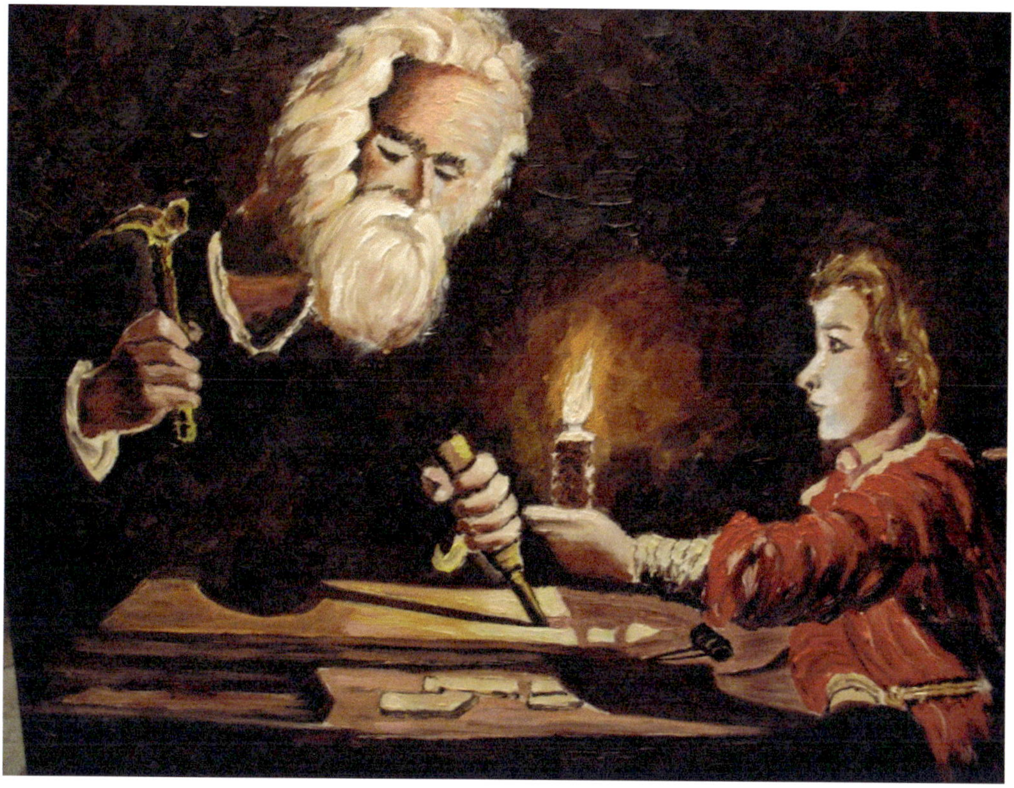

Portrait 3: "Saint Joseph: The Carpenter"
Painting by the author after Bartolomeo E. Murillo, (1650)
Oil on canvas (50X70 cm)

"Work is love made visible."

Kahlil Gibran

Some Concluding Remarks

The message that cuts across all the articles in the book is the need to work with love irrespective of the subject, trade, skill, location, space and/or time. Always we have to work with love. For only then shall we be blessed by the Master of the Universe, and only then shall we produce sweet fruits, beauty and outstanding landmarks.

As Kahlil Gibran wrote*: "Often have I heard you, as if speaking in sleep, He who works in marble and finds the shape of his own soul in the stone* (e.g., the above-mentioned world landmarks in art)*, is nobler than he who ploughs the soil* (a simple farmer)*. And he who seizes the rainbow to lay it on a cloth in the likeness of man* (e.g., Cezanne's work noted above)*, is more than he who makes the sandals for our feet. But I say, not in sleep but in the over-wakefulness of noontide, that the wind speaks not more sweetly to the giant oaks than to the least of all the blades of grass; And he alone is great who turns the voice of the wind into a song made sweeter by his own loving.*[11]*"*

He concludes*: "Work is love made visible"*. How true.

[11] Kahlil Gibran, "The Prophet", originally published in 1923, and later republished by Alfred, A Knopf, New York, in 2003.

Made in the USA
Las Vegas, NV
04 November 2026

57423061R00107